FOREST TREES AND STAND GROWTH MODELLING

Other Related Titles

Title	Author
A Manual for Dryland Afforestation and Management	G. Singh
Advances in Horticulture and Forestry (*Vol. 1-9*)	S.P. Singh
Agroforestry for Climate Resilience and Rural Livelihood	Inder Dev
Agroforestry for Sustained Productivity in Arid Regions	J.P. Gupta
Agroforestry Theory and Practices	S.B. Lal
Biodiversity Conservation	M.S.B. Kumar
Biodiversity Traditional Knowledge and Intellectual Property Rights	S. Ram Reddy
Climate Change & Himalayan Ecosystem - Indicator, Bio & Water Resources	J. Sundaresan
Climate Change & Himalayan Informatics	J. Sundaresan
Environmental Management in Mining Areas	N.C. Saxena
Essentials of Plant Nursery Management *2nd Edition*	P.K. Ray
Forest Ecology	S.S. Sagwal
Forest fires and their control	E.S. Artsybashev
Forest Mensuration Handbook	G.J. Hamilton
Forest Pathology Principles and Practice in Forestry	B.K.Bakshi
Forest Tree Seeds: Handbook	S.S. Sagwal
Forest Tribology and Anthropology	Vinod M. Mhaiske
Forestry Practicals	K.K. Chandra
Forestry Principles and Applications	A.J. Raj
Forestry Research Extension: Challenges & Strategies	C.S. Dange
Forestry Technologies- A Complete Value Chain Approach	K.T. Parthiban
Gall-Inducing Arthropods On Forest Trees	John Prasanth Jacob
Indian Manual of Plant Ecology	R. Misra
Multipurpose Trees in the Tropics: Management and Improvement Strategies	V.P. Tewari
Nursery and Plantation Practices in Forestry	Vinod Kumar
Plantation and Agroforestry: Pulpwood Value Chain Approach	K.T. Parthiban
The Community Forestry Guidebook	Jennifer Gunter
The Identification of Trees and Shrubs *2nd Edition*	F.K. Makins
Tree Breeding and Improvement: Theory and Techniques	K.T. Parthiban
Tree Mortality: Assessment And Mitigation	J.D.S. Negi
Trees on Marginal Land (Afforestation Techniques Systems)	S.S. Sagwal
Tropical Forest Ecosystems Structure & Function	V.P. Singh
Wildlife Ecology and Conservation (21st Century Biology and Agriculture: Textbook Series)	M. Balakrishnan

For more details visit website.

FOREST TREES AND STAND GROWTH MODELLING

V. P. Tewari
Former Director
Himalayan Forest Research Institute, Shimla
President, International Society of Tropical Foresters
(India Chapter)

SCIENTIFIC
PUBLISHERS

Published by

SCIENTIFIC PUBLISHERS (INDIA) 5 A, New
Pali Road, P.O. Box 91 Jodhpur 342 001 (INDIA)
E-mail: info@scientificpub.com
Website: http://www.scientificpubonline.com

Print 2026

ISBN: 978-93-91418-13-7 (HB)
ISBN: 978-93-91418-29-8 (EBook)

Printed in India

PREFACE

As with any branch of research within the natural sciences, simulation modelling and prediction are becoming increasingly important in forestry research. Growth and yield prediction models are abstract or simplified representations of some aspect of reality used primarily to estimate the future growth and yield of forest stands. Growth models can provide key information about the dynamic change of the less tangible characteristics of a forest.

Growth modelling is an essential prerequisite for evaluating the consequences of a particular management action on the future development of a forest ecosystem and is, now a day, has been central theme of Forest Management. Accurate predictions of stand growth and yield are needed for determining sustainable harvests. Technologically advanced growth and yield prediction tools can help foresters make more informed management decisions.

Since 1990s, the author has close cooperation and collaboration with the Institute of Forest Inventory and Remote Sensing, University of Göttingen, Germany and developed expertise in Forest Growth Modelling, especially with the able guidance and help from Prof. K.v. Gadow for which he will remain grateful. This book is the result of this academic exchange and learning process.

The aim of this book is to provide information about the subject to those readers who are involved in the work of forest growth modelling. The book has been written for forestry students, researchers, forestry professionals, academicians and practical foresters. Some modelling examples are included which are useful for university students, apart from the information that are added for the advanced readers.

The author gratefully acknowledges the data and information taken from various sources. Thanks are also due to the Scientific Publishers who made this publication possible. I hope this book will serve as a treasure in the collection of the students, researchers and professionals involved in forest growth modelling.

V.P. Tewari

Lucknow

ABBREVIATIONS

2SLS	two-Stage Least Squares
3SLS	three-Stage Least Squares
AAC	Allowable Annual Cut
ADA	Algebraic Difference Approach
ADE	Algebraic Difference Equation
AIC	Akaike's Information Criteria
BA	Basal Area
BAL	Basal Area of Larger trees
BALMOD	Modified Basal Area of Larger trees
CAR	Continuous Autoregressive Error
CI	Competition Index
CV	Coefficient of Variation
dbh	Diameter at breast height
FAO	Food and Agriculture Organisation
FIML	Full Information Maximum Likelihood
FNU	Fixed-area Neighbourhood Unit
FORCYTE	Forest nutrient Cycling and Yield Trend Evaluator
GADA	Generalized Algebraic Difference Approach
GMM	Generalized Method of Moments
H-H	Hansen-Hurwitz
H-T	Horvitz-Thompson
IGNP	Indira Gandhi Nahar Priyojana
LiDAR	Light Detection and Ranging
MHT	Merchantable Height
MODIS	Moderate Resolution Imaging Spectroradiometer
MRES	Mean Residual
MSE	Mean Squared Error
NASA	National Aeronautics and Space Administration
NLS	Nonlinear Least Square
NPP	Net Primary Productivity
OLS	Ordinary Least Squares
ONU	Open-area Neighbourhood Unit
PPS	Probability Proportional to Size

PRESS	Predicted Residual Sum of Squares
PSP	Permanent Sample Plot
RHS	Right-Hand Side
RMSE	Root Mean Squared Error
RS	Relative Spacing
SDMD	Stand Density Management Diagram
SEE	Standard Error of Estimate
SPOT	Système Pour l'Observation de la Terre
SRS	Simple Random Sample
SSE	Sum of Squared Error
SUR	Seemingly Unrelated Regression
UP	Primary Unit

LIST OF TABLES

LIST OF FIGURES

CONTENTS

1

Introduction

The terms "Biometrics" and "Biometry" have been used since early in the 20th century to refer to the field of development of statistical and mathematical methods applicable to data analysis problems in the biological sciences. Biometrics in reference to biological sciences has been studied and applied for several generations and is somewhat simply viewed as "biological statistics". Biometrics is the science and technology of measuring and analyzing biological data. In forestry, biometrics refers in modelling of the tree and stand development, forest data sampling and monitoring methods, forest measurements and assessments, mapping and data management technologies.

Forest biometrics is a field of study within forest and ecosystem science concerned with the analysis of quantitative problems, the development of quantitative management tools and the application of statistical methods. Forest Biometry includes (i) Data collection methods including Measurement & Mensuration, Remote Sensing, Experiments, Sampling and Inventory for the collection of tree or forest data, or data relating to processes and populations that occur within forests or trees, and (ii) Use of Statistical methods to analyze, summarize and interpret forest data. Forest measurements and biometrics involves developing and testing methods for taking and using measures from forested lands, including statistical sampling and analysis methods. Also, in today's context there is distinct applicability of biometrics in various studies concerning forest genetics, climate change modeling, quality control, design of experiments, forest fragmentation, ecological statistical techniques and vulnerability assessment.

Forest mensuration is one of the most fundamental disciplines within forest and related sciences. It deals with the measurement of trees and stands and the analysis of the resultant information. During the early days of sustained forest management, simple measurement & estimation methods and analysis of inventory & research data were available. The middle of last century, however, witnessed a worldwide increase in the need for more quantitative information about trees and stands. This generated the need for more sophisticated methods to obtain and analyse forest data. This development was followed by a phenomenal explosion of information. For a proper planning of forests, it is necessary to know the growth of individual species under varying conditions of management. Sustained Forest Management requires not only the knowledge of the existing stock of a forest but also the knowledge of the yields to be expected in future. To assess the expected future growth, it is necessary to have inventories depicting ideally periodic increments put in by a forest for the entire period of its life time.

Forest mensuration as a scientific tool originated in Europe and has always played an important role in the practice of forest management. The management of forests and tree plantations requires a quantitative estimate of the current and future volume and biomass of timber and byproducts, at national, regional and local levels. Such information is also needed for forest policy decisions. This implies the application of modern sampling concepts and sampling methods which make it possible to draw inferences about the relevant populations. Forest and forestry research require a great deal of additional information viz., about the density of forests and stands, diversity, spatial distribution of trees within stands, the size and size distribution of trees within stands and the expected growth of trees and stands (Van Laar and Akça 1997). Forest mensuration is the discipline which deals with these topics.

Forest mensuration has undergone a number of important changes in recent years. Electronic measuring devices using laser or ultrasound emissions have become commonly used to measure distances. These devices have reduced the time to measure tree heights considerably, and can be used to quickly obtain other tree parameter measures including spatial positions. Ground Positioning Systems (GPS) that triangulate satellites to determine ground positions have become widely used.

Forest management is concerned with future effects of current decisions. European normality doctorine, developed during the 19th century by Hartig, Cotta and Hundeshagen, has played an important role. The complex interaction between the level of growing stock, the timber harvest and tree growth is the guiding theme of forest management (Gadow and Bredenkamp 1992). The model of normal forest, which is also known as the fully regulated forest (Davis et al 1987, Clutter et al 1983), was first described by Hundeshagen in 1826 (Wagner 1928). The purpose of the model is to serve as a standard for comparision (Osmaston 1968) and as a tool for explaining the relationship between the growing stock, increment and felling volume (Speidel 1972). Continuety and stability of timber supply is the overridding doctorine of classical forestry science. This doctorine is expressed by the principle of sustained timber yields which aims to guarantee a constant supply of raw materials from a forest (Gadow 1990). The normality doctorine has generally featured only in theory or in working-plan policy statements. Forest management influences tree size distributions, spatial mingling of tree species and natural regeneration.

Aerial photographs are used for planning, forest inventory, and management purposes. However, these are often digital or are converted to digital formats when used in forestry today. Augmenting aerial photographs with other remotely sensed imagery has become more common. Landsat imagery has been in use for some time, but newer imagery, such as SPOT, have increased resolution both in space and in band width. LiDAR, an active sensor to measure position from an aircraft to the top of objects, is being tested for modelling crown structure and other features for inventory purposes. Other types of imagery (e.g., MODIS of NASA) have been used to estimate net primary productivity (NPP) and other productivity measures related to forest growth.

The estimation of growth is an essential step in forest management. Any planning for the future involves the prediction of future growth. Growth is a very broad and general term, meaning simply the gradual increase in a living thing by natural process. It is synonymous with the term increment in forest mensuration. The term *yield*, however, has a different meaning. Whereas growth is the amount of increase over a given period of time, yield is the total amount capable of being harvested at a given time (Spurr 1952).

Most predictions of the growth of the plant are made at one of the three levels of description. The simplest of these is the *homosphare* level where predictions are made using analogies between sites. Next is the *empirical* level where direct cause-and-effect observations are represented as two-dimensional relationships and predictions are made using those relationships. At the most detailed level of the hierarchy is the *process* level where simulations of plant physiological process are used to predict growth.

Development of sound management practices is one of the major priorities of the forestry sector. The key to successful forest management is a proper understanding of growth processes, and one of the objectives of modelling forest development is to provide the tools that enable foresters to compare alternative silvicultural treatments.

The last decades of the 18th century and the entire 19th century witnessed a relatively rapid development of forest mensurational methods. Hennert developed xylometric methods in 1791 to determine the volume of tree sections by measuring the amount of water displaced by the timber. Hennert also introduced sampling as a device to assess the volume of entire stand. Paulsen, in 1795, developed the first stem form theory and constructed the first yield tables. Cotta introduced calliper in 1804 and constructed the first volume tables. These early developments paved the way to scientific basis for forest management and forest yield studies (Van Laar and and Akça 1997). In 1948, Bitterlich introduced the revolutionary angle count concept, initially to estimate the basal area per hectare. Later on, Grosenbaugh (1952, 1958) redefined Bitterlich's variable radius method as Sampling Proportional to Size.

Reineke (1933) came out with a stand density index in even-aged stands while Yoda et al (1963) presented the self-thinning rule. Nelder (1962) gave new kinds of systematic designs for spacing experiments. A good compilation of such works may be found in the various books written on the topic viz., 'Forest mensuration' by Chapman (1921), 'Forest mensuration' by Bruce and Schumacher (1950), 'Forest inventory' by Spurr (1952), 'Forest mensuration' by Meyer (1953), 'Dendrometry' by Parde (1961), 'Forest biometrics' by Prodan (1965, 1968), 'Forest inventory' by Loetsch et al (1973),'Timber management: A quantitative approach' by Clutter et al (1983), 'Forest management' by Davis et al (1987), 'Forest management' by Gadow and Bredenkamp (1992), 'Modelling forest growth and yield: applications to mixed tropical forests' by Vanclay (1994), 'Forest measurements' by Avery and Burkhart (1994), 'Measuring trees and forests' by Philip (1994), 'Growth modelling for mixed tropical forests' by Alder (1995), 'Forest mensuration' by Van Laar and Akça (2007), 'Modelling forest development' by Gadow and Hui (1999), 'Forest mensuration' by Husch et al (2003), 'Forsteinrichtung' by Gadow (2005) and 'Modeling forest trees and stands' by Burkhart and Tomé (2012).

Enormous literature is available on forest biometrics and growth modelling as extensive research work has been conducted by the forest growth researchers covering almost each and every aspect of forest dynamics. FAO Forestry paper No. 22/2 by Alder (1980) is a very useful document for forest volume estimation and yield prediction. An analysis of growth equations has been presented by Zeide (1993). Mohren and Burkhart (1994) have presented contrasts between biologically-based process models and management-oriented growth and yield models. Forest growth models have also been evaluated by Vanclay and Skovsgaard (1997). Mendoza and Vanclay (2008) have analysed the trends in forestry modelling.

1. DATA SOURCES AND DATA STRUCTURES

A population of trees growing on a given site develops during a succession of periods of undisturbed growth and intermittent modifications of density and structure through thinning. Forest development is a direct response to various types and intensities of thinning, and is influenced by the environmental factors existing on the site. In consequence, two different kinds of empirical data are required for modelling. Firstly, data describing the change of state variables through thinning and secondly, data describing the change of state variables through natural growth. Growth data may be obtained from a variety of field experiments (Gadow and Hui 1999). The following types, which differ with regard to the dominant objective, are most common:

- *Provenance trials* for evaluating the suitability of exotic species and/or specific provenances on particular sites
- *Fertilizer trials* for investigating possible growth improvements in response to fertilizer application
- *Spacing and thinning trials,* for evaluating the effects of different planting espacements and thinning treatments on tree growth

1.1 Permanent Plots

The observations from long-term growth plots represent a very important data base for developing growth models. During long periods of time, the change of qualitative and quantitative tree attributes is assessed reiteratively in the same plot. Observations are thus obtained, permitting the construction of a growth model for a given limited set of conditions.

One of the advantages of a data base derived from permanent plots is the potential to describe polymorphic growth patterns by evaluating the data of each plot separately and by expressing the parameters of the height model as a function of site index or as a function of specific site variables. In this way, it is possible to develop non-disjoint polymorphic height models (Clutter et al 1983, Kahn 1994) and disjoint polymorphic site index equations, which depict the site-specific development of the dominant stand height over age.

A widely used model for describing the development of a variable over age is the Chapman-Richards equation, which may be expressed in the following form:

$$H = a_0 \cdot [1 - e^{-a_1 \cdot t}]^{a_2} \quad ...(1.1)$$

with, H = dominant stand height [m], t = stand age [yrs], a_0, a_1, a_2 = empirical model parameters.

By using permanent plots, it is possible to calculate a separate set of parameters for each plot. The model is *polymorphic* if the parameters determining the shape of the growth curve, a_1 and a_2, are expressed as a function of the site index (the expected height at a given reference age) or as a function of specific site variables, such as the soil depth, the geological formation or the mean annual precipitation.

A disadvantage of the permanent plot design is the high maintenance cost of the research infrastructure and the long wait for data. The object of the trial is not always achieved, as plots may be destroyed prematurely by wind or fire, or by unauthorized cutting.

1.2 Temporary Plots

Temporary plots may provide a quick solution in a situation where nothing is known about forest development. They are measured only once, but cover a wide range of ages and growing sites. Thus, the sequence of re-measurements in time is substituted by simultaneous *point* measurements in space. This method has been used extensively during the 19th century (Kramer 1988, p. 97, Assmann 1953, Wenk et al 1990).

The main limitation of temporary plots is the fact that they cannot provide information about the rate of change of a known initial state. Thus, some of the more effective contemporary techniques of growth modelling using a system of differential equations cannot be applied (Garcia 1988).

1.3 Interval Plots

A compromise may be achieved by using a system of growth trials which maintains the advantages of permanent plots, i.e., obtaining rates of change of known initial states, as well as temporary plots, i.e., broad coverage of initial states and minimum delay until suitable data are available.

Interval plots are measured twice, the interval between the measurements being sufficiently long to absorb the short-term effects of abnormal climatic extremes. In most cases, an interval of 5 years would be appropriate. The interval is a period of undisturbed growth. Silvicultural operations are not permitted to take place between the two measurements. Measurements should coincide with a thinning operation, to obtain data not only about tree growth, but at the same time about the change of state variables resulting from a thinning. The thinning effects may be assessed at the initial (t_1) or at the final (t_2) measurement, or at both occasions.

A suitable model type utilizing the data from two successive measurements is the *algebraic difference form of a growth function*, which has been used by a number of authors (e.g., Clutter et al 1983, Ramirez-Maldonado et al 1988, Forss et al 1996).

Interval plots, measured twice and spread over a range of growing sites, development stages and silvicultural treatment categories, combine the advantages of the permanent and the temporary designs. Thus, the rates of change of known initial states can be assessed and, after minimum delay, suitable data are available for modelling.

The interval concept offers flexibility, as the plots may be abandoned at any time after achieving the objective to obtain one interval measurement. A permanent infrastructure involving staff for repeated re-measurements is thus not required.

2. PLOT SIZE - VARIABILITY AND OPTIMISATION

Knowledge of variability among sampling units is essential in designing an efficient sample survey. The information can sometimes be obtained from previous surveys of the same or similar populations. When this is done, the effect of plot size on variability must be considered. The only sure way of knowing how variability changes with plot size is to make a special study of the population in question (Freese 1961). If that does not seem warranted, an approximation is to assume that the squared coefficient of variation $(CV)^2$ is inversely proportional to the square root of the relative plot size (P). Thus, if the coefficient of variation (CV_1) for plots of size P_1 is known, the estimated coefficient (CV_2) for plots of the same shape and orientation but different size (P_2) would be:

$$(CV_2)^2 = (CV_1)^2 \sqrt{(P_1/P_2)} \qquad ...(1.2)$$

Optimisation of plot size is an important practical problem in forest inventory. The optimal plot size is that which minimises total time required for location and measurement for a stated accuracy in the desired variable. Thus, the problem is (1) to express total time as a function of plot size and (2) to find the plot size which corresponds to the minimum of this function (Zeide 1980).

2.1 Total time

Total time T to measure n_i plots of size P_i in order to get desired level of accuracy E can be presented as follows:

$$T = n_i(t_i + m_i) \qquad ...(1.3)$$

where, t_i = travel time per plot of size P_i and, m_i = plot-measurement time for one plot of size P_i.

Given a desired accuracy E and a specified confidence interval, the required number of plots is defined by:

$$n_i = \left(\frac{C_i t}{E}\right)^2 \qquad ...(1.4)$$

where, C_i = coefficient of variation between plots of size P_i

$E = \frac{ts_{\bar{x}}}{\bar{x}}$ = desired accuracy, $s_{\bar{x}}$ =standard error of mean $\bar{x}$, t = appropriate value of Student's t-statistic.

Using Freese's approximation, the number of plots as a function of plot size may be expressed as:

$$n_i = \left(\frac{C_i t}{E}\right)^2 = \frac{C^2 P^{\frac{1}{2}} t^2}{E^2} P_i^{\frac{-1}{2}} = kP_i^{\frac{-1}{2}} \quad ...(1.5)$$

$$\text{where, } k = \frac{C^2 P^{\frac{1}{2}} t^2}{E^2} = P_n^{\frac{1}{2}}$$

2.2 Travel Time as a Function of Plot Size

Travel time between two plots (t_i) is usually proportional to the distance between the plots and inversely proportional to travel speed. For systematic design when n_i points are evenly distributed on a square lattice (which is most common practical design) the distance between two nearest points can be calculated as follows (Zeide 1980):

$$\text{Distance} = \sqrt{(\text{total area}/n_i)} \quad ...(1.6)$$

The same is true for the average minimum distance between points in random sampling (O'Regan and Arvanitis 1966).

Hence, we can write:

$$t_i = (\sqrt{\text{total area}})/\text{travel speed} \times n_i^{-1/2} = aP_i^{1/4} \quad ...(1.7)$$

where, $a = \sqrt{(\text{total area}/k)} \times (1/\text{travel speed})$

2.3 Plot Measurement Time as a Function of Plot Size

Plot measurement time (m_i) depends first of all on the number of trees to be tallied. Accordingly, plot-measurement time is directly proportional to number of trees and, hence, to plot size (Zeide 1980).

In addition to tree number, two other factors must be considered. We tend to slow down the tally unconsciously near the plot and sector boundaries as well as to check consciously the location of some boundary trees. The boundary area of a certain width is proportional to the plot radius, not plot size (in the case of square plot, boundary area is proportional to side length). The third factor is the time required to lay out the plot and mark boundaries. Thus, two factors of plot measurement are proportional to the plot radius, i.e., to the square root of the plot size, while the first mentioned factor is proportional to the plot size. Hence, we can assume that plot-measurement time is proportional to the plot size raised to the power c which lies somewhere between ½ and 1.

$$M_i = bP_i^c \quad ...(1.8)$$

where, b = coefficient of proportionality, and c = exponent of plot size ($\frac{1}{2} < c < 1$).

2.4 Optimisation of Plot Size

Total time can be rewritten now as a function of plot size:

$$T = kP_i^{-1/2} (aP_i^{1/4} + bP_i^{3/4}) \quad ...(1.9)$$

Plot size corresponding to the minimum total time can be found from the following equation:

$$dT/dP = k\,[-(\tfrac{1}{4})\,aP_i^{-5/4} + (\tfrac{1}{4})\,bP_i^{-3/4}] = 0 \qquad ...(1.10)$$

To tell whether the extremum is the minimum, we take the second derivative of the total time function:

$$D^2T/dP^2 = k[5/4\,P_i^{-1}\,(\tfrac{1}{4}\,aP_i^{-5/4}) - 3/4\,P_i^{-1}\,(\tfrac{1}{4}\,bP_i^{-3/4})] \qquad ...(1.11)$$

From the function dT/dP, it follows that:

$$\tfrac{1}{4}\,aP_i^{-5/4} = \tfrac{1}{4}\,bP_i^{-3/4} = H > 0 \qquad ...(1.12)$$

Hence,

$$D^2T/dP^2 = \mathrm{kH}\,(5/4\,P_i^{-1} - 3/4\,P_i^{-1}) = \tfrac{1}{2}\,\mathrm{kH}\,P_i^{-1} > 0 \qquad ...(1.13)$$

Since the second derivative is positive the extremum is minimum. Therefore:

Optimal plot size $= P^* = (a/b)^2$ ...(1.14)

2.5 Application

It is convenient to express the coefficients a and b in terms of travel and plot-measurement time for plots of a certain size P. We can write:

$$a = tP^{-1/4} \qquad ...(1.15)$$

$$b = mP^{-3/4} \qquad ...(1.16)$$

Therefore,

$$P^* = (tP^{-1/4}/mP^{-3/4})^2 = P\,(t/m)^2 \qquad ...(1\text{-}17)$$

Thus, to calculate optimal plot size one must first lay out a few plots of certain size. The plots should be situated in different parts of a stand to avoid the effect of patchiness. Travel speed between plots and plot-measurement time, m, should be recorded. The second step is to calculate the coefficient of variation between the plots, C, and the number of plots, n, required for a given accuracy E. Third, one finds the distance between neighbouring plots and travel time between them, t, dividing the distance by travel speed. Next one calculates the optimal plot size (Zeide 1980). Applying Freese's equation, one can calculate variation between plots of optimal size and find number of them which is necessary to get same accuracy E. When one uses this number and the optimal plot size the total time to get a given accuracy E will be minimal.

3. FIXED-RADIUS PLOTS OR VARIABLE-RADIUS PLOTS?

Sampling with fixed-radius plots is a special, restricted variant of point-sampling with variable radii. Therefore, there are no theoretical advantages of the fixed-radius over the more general variable-radius alternative. Rather, sampling with variable radii offers greater flexibility for achieving better biological representation of stand conditions and increased statistical efficiency. However, sampling with variable radii may demand more careful field work. Problems of field procedure can be reduced by designs that combine fixed-radius sampling rules for the smallest and largest dbh classes with variable-radius rules for intermediate dbh classes (Stage and Rennie 1994). If the width of the smallest class is broad enough that all ingrowth into the population will be sampled at subsequent examinations by the fixed-radius rule, then most of the perceived difficulties of calculating increment from remeasured plots can be avoided.

4. FOREST GROWTH MODELS

As with any branch of research within the natural sciences, simulation modeling and prediction are becoming increasingly important in forestry research. Development of sound management practices is one of the major priorities of the forestry sector. Growth and yield prediction models are abstract or simplified representations of some aspect of reality used primarily to estimate the future growth and yield of forest stands. Growth models can provide key information about the dynamic change of the less tangible characteristics of a forest, such as the *stability* and *resilience* in an environment affected by industrial pollution or the aesthetic value of a given forest structure (Gadow and Hui 1999).

Growth modelling is an essential prerequisite for evaluating the consequences of a particular management action on the future development of a forest ecosystem and is, now a day, has been central theme of Forest Management. Accurate predictions of stand growth and yield are needed for determining sustainable harvests. Technologically advanced growth and yield prediction tools can help foresters make more informed management decisions. Efficiency of the forest production system is to be augmented by optimal management decisions, which need quality information on basic parameters of forests. The availability of such information is scanty and concern has been expressed about its inadequacy. Therefore, growth & yield studies based on the permanent sample plot (PSP) data are required for the sustainable management of the forests and plantations.

As more data become available, the modelling approach evolves to produce sometimes elaborate models viz., whole stand model, diameter distribution model and single tree model. Provided the conditions under which these models are applied are the same as those under which the basic data have been collected, such models provide robust tools for prediction of future yield. They can also be used for the evaluation of stand structure and stand productivity in a particular field condition and in general for decision support in forest management and planning. Management oriented empirical growth and yield models are very much relevant in transferring of summary information on forest growth to the forest managers.

The most appropriate modelling technique is determined by the level of detail of the available data and by the level of resolution of the required forecast. Thus, a stand model will be used if average population values and area-based information such as the basal area is available. An individual tree model is appropriate when the relevant size attributes of a tree and the attributes and coordinates of its immediate neighbours are available (Gadow and Hui 1999).

A stand growth model represents an abstraction of the natural dynamics of a forest stand, and depicts growth, mortality and other changes in stand composition and structure. It also mathematically describes the growth and yield of trees and stands.

Some models are developed to predict Yield, which is the final accumulated growth at the end of a certain period (e.g., total volume growth in cubic meters per hectare); while others predict Growth, which is the total increase in dimensions of one or more individuals in a forest stand over a given period of time (e.g., total volume growth in cubic meters per hectare and per year), as well as Yield.

Common usage of the 'Growth model' generally refers to a system of equations which can predict the growth and yield of forest stand under wide variety of conditions. Thus, a growth model may comprise a series of mathematical equations, the numerical values embedded in those equations, the logic necessary to link these equations in a meaningful way. In its broadest sense, the term may also embrace yield tables and curves, which are analogous to equation, but which have been stated in a tabular or graphical form rather than a mathematical form.

Most growth models are developed on the basis of empirical data collected in trials of varying design. Permanent growth series, which are rather common in Central Europe, have been and are being observed and re-measured over long periods of time. Temporary plots, distributed over a wide range of ages and growing sites, are sometimes used to avoid the long wait, which is a disadvantage of the permanent growth series. A compromise may be found by using *interval plots,* which are also spread over a wide range of initial system states, but are measured only twice to include one growth interval (Gadow and Hui 1999).

4.1 Types of Growth Models

Forest models represent average experience of how trees grow and of how forest structures are modified. The level of detail of these models differs greatly. Tree models deal with morphological details of branching, stem form and root growth. Regional production models and stand growth models produce aggregate information about the development of a population of trees with a given set of environmental conditions and given intermittent modifications of stand attributes through human interference and other disturbances. The mixed effect model is used for modelling hierarchical populations, consisting of groups and individuals within groups. For example, the group may be a sample plot, and the individual may be the tree within the plot.

Among the common types of growth models are highly aggregated volume-over-age models for regional yield forecasting, stand models for predicting basal area growth as a function of age and initial basal area, distance-independent size class models for predicting the movement of diameter distributions and distance-dependent individual tree models requiring spatial information in two or three dimensions. These models are further subdivided according to how the stand density is modelled. For instance, variable density whole stand models can assess the effects of yield on variation in stand density (e.g., crown cover, basal area). Of the individual tree models, only distance-dependent models maintain a spatial record of the point density around the individual trees (Munro 1974).

Regional yield models are represented by highly aggregated yield-over-age equations. They are used in resource forecasting, specifically for predicting the development of a given age-class distribution in response to a series of periodic harvest levels. *Stand growth models* require more information and, in turn, provide a greater degree of detail, including estimates of dominant height, stand average diameter, volume per hectare, basal area and stems per hectare. The basic modelling unit in a *size class model* is a representative tree impersonating a number of trees within a size class or cohort. Size class models, requiring even more detailed information than stand models, are probably the most common type for simulating alternative silvicultural programs. *Individual tree models* utilize information about the position and size of specific trees (e.g., tree heights, diameters and crown lengths) and of the trees in their immediate neighbourhood (Gadow and Hui 1999). The spatial information may be available in two or three dimensions. Three dimensional spatial models are used to quantify the amount of shading and constriction of the growing space, caused by neighbouring trees.

Two other classes of models are Process model and Succession model. Process models attempt to model the processes of growth, taking as input the light, temperature and soil nutrient levels, and modelling photosynthesis, respiration and allocation of photosynthates to roots, stems and leaves (e.g., Landberg 1986, Mäkelä 1992). These are also known as mechanistic or physiological models. These models help to provide a better understanding of growth and stand dynamics, but have not yet successfully been used for predicting timber yield for forest management. Ecosystem succession models (e.g., Shugart 1984, Botkin 1993) attempt to model species succession, but are generally unable to provide reliable information on timber yield. The FORCYTE model (Kimminis 1988) uses a hybrid approach and can predict forest yields under a variety of nutrient regimes.

It is useful to distinguish between models for understanding and models for prediction. Models for understanding (e.g., process models) are useful to comprehend and link previously isolated bits of knowledge and may help to identify gaps where more work is needed. The benefits come from the insights gained while developing and exploring the model, and future uses are less important. Conversely, models for prediction may sacrifice specific details of growth processes to achieve greater efficiency and accuracy in providing information for forest management. Realism is not necessarily a virtue in a model, and it may be better to abstract just those aspects that are most relevant in each instance.

Finally, irrespective of its detail, a model may be deterministic or stochastic. A deterministic growth model gives an estimate of the expected growth of a forest stand, in the same way that the mean indicates the expected trend for a population. Given the same initial conditions, a deterministic model will always predict the same result. However, because of natural variation in the environment, real forest stands may not grow exactly the same amount each year, but may grow more or less than expected amount (Gadow and Hui 1999).

A stochastic model attempts to illustrate this natural variation by providing different predictions, each with a specific probability of occurrence; any one of these estimates may correspond exactly to the growth under some circumstances, but may differ from the expected growth. A single estimate from a stochastic model is of little use, as a whole series of estimates is necessary to provide useful information of the variability of predictions. For instance, twenty estimates from a stochastic model will not only give a good indication of the expected growth (the mean or median), but also of its variability. The term 'stochastic' is used in preference to the term 'probabilistic', which is sometimes used in other contexts.

Deterministic and stochastic models serve complementary purposes. Deterministic models are effective for determining the expected yield, and may be used to indicate the optimum stand condition. Stochastic models may indicate the reliability of these predictions, and the risks associated with any particular regime. Both deterministic and stochastic predictions can be obtained from some models (Gadow and Hui 1999). Although stochastic models can provide some useful information not available from deterministic models, most of the information needs for forest planner and managers can be provided more efficiently with deterministic models.

4.2 Caution About Models

Forestry and statistically-based biological experimentation are both relatively new sciences whose joint development is governed largely by the (slow) rate of tree growth. Seeming contradictions among the limited existing experiments serve to highlight our imperfect understanding of complex biological systems and discourage risk-laden investment decisions based on limited (or select) information. Decision making given imperfect information requires risk analyses which take into account the uncertainties regarding future biological and economic consequences. Models can be important tools, but one should not rely solely on them for making decisions. Professional judgment should be used to examine the data and assumptions before making the final management decision.

Each model has a unique niche; no one model is applicable in all situations. To get the most benefit from these models, foresters have to understand the basic differences among them to select the appropriate model for a given situation. Differences among models stem from differences in the databases used to calibrate them and differences in model architecture. Perfect databases do not exist. Data quality and quantity are always in short supply since funds are limited and long-term growth data requires time (www.for.gov.bc.ca).

Likewise, there is no one "best" modelling approach. A model's architecture stems from the modelling approach chosen by the modeller based on the intended application and available data. Limited databases along with the limited knowledge of tree and stand growth necessarily lead to different approaches for different needs and applications. For instance, a primary emphasis on supporting silvicultural prescriptions is likely to lead to a different model than an emphasis on inventory or planning. Similarly, an emphasis on single-rotation yields will produce a different model than an emphasis on long-term sustainability.

4.3 Application and uses of Model

Proper use of a model depends on proper selection and preparation of the input data and proper interpretation of the model output. The main uses of growth & yield predictions are to:

- increment and update forest inventories
- compare silviculture treatments by simulating treatments and predicting outcomes
- influence stand and forest level decision making
- provide input for forest management planning including timber supply analysis, Allowable Annual Cut (AAC) determinations and policy making
- assess the impact of timber losses due to pests and fire
- allow extrapolations for missing or inadequate data
- explore and teach tree and stand dynamics

The application of any model in silvicultural decision support also requires a clear statement of management objectives translated into appropriate quantitative values that can be identified in model output. Care must be taken to understand the implications and limitations of using various quantitative measures as surrogates for management objectives.

For instance, the common use of average diameter as a surrogate for piece size is easily misinterpreted when exploring thinning response. The immediate shift in diameter distribution (and average diameter) due to the thinning operation itself (alias "the chain saw effect") confounds later interpretations of average diameter and thinning response. Thinning from below immediately narrows the diameter distribution and increases the average diameter. This in turn exaggerates the effect of thinning on the time required to reach a certain piece-size. Examining the largest diameter classes via stand and stock tables, provided by some models, gives a clearer picture of thinning response than average diameter does (www.for.gov.bc.ca).

Growth models are of limited use on their own, and require ancillary data to provide useful information. With suitable inventory and other resources data; growth models provide a reliable way to examine silvicultural and harvesting options; to determine the sustainable timber yield; and examine the impacts of forest management and harvesting on other values of the forest. Growth models may also have a broader role in forest management and in the formulation of forest policy. Used to advantage and in conjunction with other resource and environmental data, growth models can be used to make predictions, formulate prescriptions and guide forest policy.

Models are to be developed which would be best suited for particular species, under specific management situations. To be effective, the developed models should be statistically sound, bio-mathematically realistic and perform as efficient predictive tools. Indeed, models are projections for the future, and none can represent real life perfectly. Nevertheless, they are useful, as the decision-making process often needs to know the future unknown. The models must also be validated with an independent data set for their wider applicability.

2

Sampling Techniques and Experimental Design

1. SAMPLING TECHNIQUES

In statistics, quality assurance, and survey methodology, sampling is concerned with the selection of a subset of individuals from within a statistical population to estimate characteristics of the whole population. Acceptance sampling is used to determine if a production lot of material meets the governing specifications. Two advantages of sampling are that the cost is lower and data collection is faster than measuring the entire population.

Each observation measures one or more properties of observable bodies distinguished as independent objects or individuals. In survey sampling, weights can be applied to the data to adjust for the sample design, particularly stratified sampling (blocking). Results from probability theory and statistical theory are employed to guide practice.

An enumeration of the entire tree population is extremely rare. Sampling from a population of tree characteristics is the common approach today. The theory and practice of tree sampling embraces a great variety of methods. Good sampling is characterized by unbiasedness, a small sampling error and consistency. Unbiasedness implies that the mathematical expectation of the sample statistic is identical to the population parameter. An estimate is consistent if, with increasing sample size, the sample value approaches the true population value (Gadow 1990).

1.1 Probability Sampling

A probability sampling is one in which every unit in the population has a chance (greater than zero) of being selected in the sample, and this probability can be accurately determined. The combination of these traits makes it possible to produce unbiased estimates of population totals, by weighting sampled units according to their probability of selection.

Probability sampling includes: Simple Random Sampling, Systematic Sampling, Stratified Sampling, Probability Proportional to Size sampling, and Cluster or Multistage Sampling. These various ways of probability sampling have two things in common: (1) every element has a known nonzero probability of being sampled, (2) involves random selection at some point.

1.1.1 Simple Random Sampling

In a simple random sample (SRS) of a given size, all such subsets of the frame are given an equal probability. Each member of the population thus has the same probability of being sampled. Furthermore, any given pair of elements has the same chance of selection as any other such pair. This minimizes bias and simplifies analysis of results. In particular, the variance between individual results within the sample is a good indicator of variance in the overall population, which makes it relatively easy to estimate the accuracy of results.

However, SRS can be vulnerable to sampling error because the randomness of the selection may result in a sample that doesn‘t reflect the makeup of the population. Systematic and stratified techniques attempt to overcome this problem by using information about the population to choose a more representative sample.

SRS may also be cumbersome and tedious when sampling from an unusually large target population. In some cases, investigators are interested in research questions specific to subgroups of the population. SRS cannot accommodate the needs of researchers in this situation because it does not provide subsamples of the population. Stratified sampling addresses this weakness of SRS.

1.1.2 Systematic Sampling

Systematic sampling relies on arranging the target population according to some ordering scheme and then selecting elements at regular intervals through that ordered list. Systematic sampling involves a random start and then proceeds with the selection of every *k*th element from then onwards. In this case, k = population size/sample size. It is important that the starting point is not automatically the first in the list, but is instead randomly chosen from within the first to the *k*th element in the list.

As long as the starting point is randomized, systematic sampling is a type of probability sampling. It is easy to implement and the stratification induced can make it efficient, if the variable by which the list is ordered is correlated with the variable of interest.

However, systematic sampling is especially vulnerable to periodicities in the list. If periodicity is present and the period is a multiple or factor of the interval used, the sample is especially likely to be unrepresentative of the overall population, making the scheme less accurate than simple random sampling. Another drawback of systematic sampling is that even in scenarios where it is more accurate than SRS, its theoretical properties make it difficult to quantify that accuracy.

1.1.3 Stratified Sampling

Where the population embraces a number of distinct categories, the population is subdivided into homogeneous classes known as "strata". Each stratum is then sampled as an independent sub-population, out of which individual elements can be randomly selected (Groves et al 2004). There are several potential benefits to stratified sampling.

First, dividing the population into distinct, independent strata can enable researchers to draw inferences about specific subgroups that may be lost in a more generalized random sample. Second, utilizing a stratified sampling method can lead to more efficient statistical estimates (provided that strata are selected based upon relevance to the criterion in question, instead of availability of the samples). Even if a stratified sampling approach does not lead to increased statistical efficiency, such a tactic will not result in less efficiency than would simple random sampling, provided that each stratum is proportional to the group‘s size in the population. Third, it is sometimes the case that data are more readily available for individual, pre-existing strata within a population than for the overall population; in such cases, using a stratified sampling approach may be more convenient than aggregating data across groups (though this may potentially be at odds with the previously noted importance of utilizing criterion-relevant strata). Finally, since each stratum is treated as an independent population, different sampling approaches can be applied to different strata, potentially enabling researchers to use the approach best suited (or most cost-effective) for each identified subgroup within the population.

There are, however, some potential drawbacks to using stratified sampling. First, identifying strata and implementing such an approach can increase the cost and complexity of sample selection, as well as leading to increased complexity of population estimates. Second, when examining multiple criteria, stratifying variables may be related to some, but not to others, further complicating the design, and potentially reducing the utility of the strata. Finally, in some cases (such as designs with a large number of strata, or those with a specified minimum sample size per group), stratified sampling can potentially require a larger sample than would other methods (although in most cases, the required sample size would be no larger than would be required for simple random sampling).

A stratified sampling approach is most effective if variability within strata are minimized, variability between strata are maximized and the variables upon which the population is stratified are strongly correlated with the desired dependent variable.

Advantages:

- Focuses on important subpopulations and ignores irrelevant ones
- Allows use of different sampling techniques, for different subpopulations
- Improves the accuracy/efficiency of estimation
- Permits greater balancing of statistical power of tests of differences between strata by sampling equal numbers from strata varying widely in size

Disadvantages:

- Requires selection of relevant stratification variables which can be difficult
- Is not useful when there are no homogeneous subgroups
- Can be expensive to implement

1.1.3.1 Post-stratification

Stratification is sometimes introduced after the sampling phase in a process called "post-stratification" (Groves et al 2004). This approach is typically implemented due to a lack of prior knowledge of an appropriate stratifying variable or when the experimenter lacks the necessary information to create a stratifying variable during the sampling phase. Although the method is susceptible to the pitfalls of post hoc approaches, it can provide several benefits in the right situation. Implementation usually follows a simple random sample. In addition to allowing for stratification on an ancillary variable, post-stratification can be used to implement weighting, which can improve the precision of a sample‘s estimates (Groves et al 2004).

1.1.4 Cluster Sampling

Sometimes it is more cost-effective to select respondents in groups ('clusters'). Sampling is often clustered by geography, or by time periods. Clustering can reduce travel and administrative costs. Cluster sampling is used when the cost of locating sampling units is high relative to the cost of taking measurements. The sampling unit is a cluster of individuals and the unit value is the proportion of the individuals in the group having the specified attribute.

Cluster sampling generally increases the variability of sample estimates above that of simple random sampling, depending on how the clusters differ between themselves, as compared with the within-cluster variation. For this reason, cluster sampling requires a larger sample than SRS to achieve the same level of accuracy - but cost savings from clustering might still make this a cheaper option.

Cluster sampling is commonly implemented as multistage sampling. This is a complex form of cluster sampling in which two or more levels of units are embedded one in the other. The first stage consists of constructing the clusters that will be used to sample from. In the second stage, a sample of primary units is randomly selected from each cluster (rather than using all units contained in all selected clusters). In following stages, in each of those selected clusters, additional samples of units are selected, and so on. All ultimate units (individuals, for instance) selected at the last step of this procedure are then surveyed. This technique, thus, is essentially the process of taking random subsamples of preceding random samples.

1.1.5 Multistage Sampling

It is sometimes practical to concentrate the sampling effort to reduce walking time. Sampling is usually done in two stages. The first stage involves drawing at random m out of M primary units (UP) of equal size. At the second stage within each UP, at random n out of N possible secondary units (US) are selected (Gadow and Bredenkamp 1992).

Multistage sampling can substantially reduce sampling costs, where the complete population list would need to be constructed. By eliminating the work involved in describing clusters that are not selected, multistage sampling can reduce the large costs associated with traditional cluster sampling (Lohr 1999).

1.1.6 Probability-proportional-to-size Sampling

In some cases, the sample designer has access to an "auxiliary variable" or "size measure", believed to be correlated to the variable of interest, for each element in the population. These data can be used to improve accuracy in sample design. One option is to use probability-proportional-to-size (PPS) sampling, in which the selection probability for each element is set to be proportional to its size measure, up to a maximum of one. In a simple PPS design, these selection probabilities can then be used as the basis for Poisson sampling. However, this has the drawback of variable sample size, and different portions of the population may still be over- or under-represented due to chance variation in selections. To address this problem, PPS may be combined with a systematic approach.

The PPS approach can improve accuracy for a given sample size by concentrating sample on large elements that have the greatest impact on population estimates. In some cases, an older measurement of the variable of interest can be used as an auxiliary variable when attempting to produce more current estimates (Lohr 1999).

1.2 Unequal Probability Sampling

In simple random sampling, the probability that each unit will be sampled is the same. Sometimes, estimates can be improved by varying the probabilities with which units are sampled. The sampling error may be reduced if different probabilities of being selected are assigned to sampling units. Unequal probability sampling is when some units in the population have different inclusion probabilities from others. The Hanson-Hurwitz (H-H) estimator and Horvitz-Thompson (H-T) estimators are used in this sampling.

1.2.1 The Hanson-Hurwitz and Horvitz-Thompson Estimator

The Hansen-Hurwitz estimator gives the framework for all unequal probability sampling with replacement (Hansen and Hurwitz 1943). "With replacement" means that the selection probabilities are the same for all draws; if selected elements would not be replaced (put back to the population), the selection probabilities would change after each draw for the remaining elements. Suppose a sample of size n is selected randomly with replacement from a population, but that on each draw, unit *i* has probability *pi* of being selected, so that $\sum_{i=1}^{N} p_i = 1$.

The probability p_i is called the selection probability for the ith unit. If the response variable is positively correlated with the selection probability, then the Hansen-Hurwitz estimator can have lower variance than the estimator based on an SRS.

The Horvitz-Thompson estimator is a general estimator for a population total, which can be used for any probability sampling plan. This includes both sampling with and without replacement. The higher the probability of selection, π_i, of a unit i to the sample, the less weight the corresponding response y_i is given. In this way the *H-T* estimator, like the Hansen-Hurwitz estimator, uses probability to weight the responses in estimating the total.

The primary difference between the *H-T* and *H-H* estimator is the fact that the former uses the inclusion probability (π_i) of the units to the sample, whereas the latter uses the probability of selection (p_i) of a unit for a single draw. The *H-H* estimator is restricted to random sampling with replacement while the H-T estimator can be used in much wider range of sampling plans.

1.3 Spatial Sampling

The main aim of spatial sampling is to collect samples in 1-, 2- or 3-dimensional space. It is typically used to estimate the total or mean for a parameter in an area, to optimize parameter estimations for un-sampled locations, or to predict the location of a movable object. Some objectives are for populations, representing the "here and now", whereas other objectives concern super-populations that generate the populations. Data to be collected are usually spatially auto-correlated and heterogeneous, whereas sampling is usually not repeatable. In various senses it is distinct from the assumption of independent and identically distributed (i.i.d.) data from a population in conventional sampling.

The uncertainty for spatial sample estimation propagates along a chain from spatial variation in the stochastic field to sample distribution and statistical tools used to obtain an estimate. This uncertainty is measured using either a design-based or model-based method. Both methods can be used in population and super-population studies. An unbiased estimate with the lowest variance is thus a common goal in spatial sampling and inference. Reaching this objective can be addressed by sample allocation in an area to obtain a restricted objective function (Wang et al 2012).

1.4 Ratio/Regression Estimator

For the ratio estimator and the regression estimator, it is stipulated that the parametric mean or the true total of the ancillary variable needs to be known in order to apply those estimators. In some cases, this is a very unpleasant situation because the population values might not be known. A way out is to also estimate these values. This is exactly what double sampling is about, also referred to as two-phase sampling: in a first phase, the ancillary variable is estimated, usually with a relatively large sample of a variable that is relatively easy and inexpensive to observe. Then, in a second phase, a smaller sample is taken of the target variable, which is frequently a variable much more expensive or difficult to observe; simultaneously, however, also the ancillary variable is observed, so that a relationship between target and ancillary variable can be established (either a ratio in the case of double sampling with the ratio estimator or a regression in the case of double sampling with the regression estimator). Here, the correlation to the ancillary variable is also used to reduce the sample size in the second phase (Klienn 2007).

There are situations in forest inventory sampling in which the value of the target variable is known (or suspected) to be highly correlated to another variable (called co-variable or ancillary variable). We know from study of spatial autocorrelation and the optimization of cluster plot design that correlation between two variables means essentially that the one variable does yet contain a certain amount of information about the other variable. The higher the correlation the better can the value of the second variable be predicted when the value of the first is known (and vice versa). Therefore, if such a co-variable is there on the plot, it would make sense to also observe it and utilize the correlation to the target variable to eventually improve the precision of estimating the target variable. This is exactly the situation where the ratio estimator is applied.

A typical and basic example is that of sample plots of unequal size. Another example would be clusters of unequal size: while cluster sampling only deals with the case of clusters of equal size, the ratio estimator allows taking into account the differently sized sample plots. Oviously, the size of the sample plots is the co-variable that needs to be determined for each sample plot. A high positive correlation is to be expected between most forestry-relevant attributes and plot size should be obvious. The larger the plot the more basal area, number of stems, volume etc. is expected to be present.

In fact, the ratio estimator does not introduce a new sampling technique. The ratio estimator introduces new elements

- into the plot design (observation of an ancillary variable on each plot, in addition to the observation of the target variable); and above all
- into the estimation design; the ratio estimator integrates the ancillary variable into the estimator

The general procedure for both double sampling with the ratio estimator and for double sampling with the regression estimator is identical. Contrary to double sampling for stratification where a categorical variable is observed in the first phase, it is usually metric variables that serve as ancillary variables when double sampling with the ratio or regression estimator is being used.

In the first phase, a sample of size 'n' is taken to estimate the mean or total of the ancillary variable X. The sample taken is usually large because measurement of X is cheap, fast and easy. In the second phase, a sample is selected on which both target and ancillary variable are observed; from these pairs of observations, a relationship between the two variables can be established, either a ratio or a regression. The second phase sample is usually small because the observation of Y is usually more expensive, difficult and time consuming. Then, the observations from the first phase are used to estimate the total and mean of the target variable for the entire area of interest.

In both approaches, dependent or independent phases are possible and the corresponding estimators need to be used (Klienn 2007). It is interesting to note, that double sampling is also interesting in context of Sampling with partial replacement (SPR) that is a very efficient technique to estimate changes.

1.5 Sample Size Determination and Allocation

The projected sample size estimate is usually based on the Coefficient of variations (CVs) of major variables. The allocation of the sample size across domains may be equal allocation, proportional allocation, square root allocation and Kish Allocation.

In case of heterogeneity, larger sample size is needed to study more diverse population. Also, larger sample size is needed to get smaller error. Design Effects will be smaller if stratified and larger if cluster. Complex multivariate statistics need larger samples. Accuracy of sample depends upon sample size and not on the ratio of sample to population.

For sample size determination, the degree of precision, $SE(z)$ required need to be specified. CV determine the sample size needed to give the specified level of precision

$$CV(z) = SE(z)/Z \quad \text{...(2.1)}$$

Sample Size is determined at the domain. If there is stratification within a domain then sample size must be allocated across the strata. Sample size determination is the act of choosing the number of observations or replicates to include in a statistical sample. The sample size is an important feature of any empirical study in which the goal is to make inferences about a population from a sample. In practice, the sample size used in a study is determined based on the expense of data collection, and the need to have sufficient statistical power. In complicated studies there may be several different sample sizes involved in the study: for example, in a survey sampling involving stratified sampling there would be different sample sizes for each population. Larger sample sizes generally lead to increased precision when estimating unknown parameters.

2. DESIGN OF EXPERIMENTS

A methodology for designing experiments was proposed by Fisher (1935). Much of his pioneering work dealt with agricultural applications of statistical methods. These methods have been broadly adapted in the physical and social sciences. In some fields of study, it is not possible to have independent measurements to a traceable standard. Comparisons between treatments are much more valuable and are usually preferable. Often one compares against a scientific control or traditional treatment that acts as baseline. Randomization, blocking and replications are used in experimental design to minimize variance. Factorial designs are used where there are number of independent variables.

Random assignment is the process of assigning individuals at random to groups or to different groups in an experiment. The random assignment of individuals to groups (or conditions within a group) distinguishes a rigorous, "true“ experiment from an adequate, but less-than-rigorous, "quasi-experiment" (Creswell 2008). Provided the sample size is adequate, the risks associated with random allocation (such as failing to obtain a representative sample in a survey, or having a serious imbalance in a key characteristic between a treatment group and a control group) are calculable and hence can be managed down to an acceptable level. Random does not mean haphazard, and great care must be taken that appropriate random methods are used.

Measurements are usually subject to variation and uncertainty. Measurements are repeated and full experiments are replicated to help identify the sources of variation, to better estimate the true effects of treatments, to further strengthen the experiment's reliability and validity, and to add to the existing knowledge of the topic. However, certain conditions must be met before the replication of the experiment is commenced (Burman 2010).

Blocking is the arrangement of experimental units into groups (blocks) consisting of units that are similar to one another. Blocking reduces known but irrelevant sources of variation between units and thus allows greater precision in the estimation of the source of variation under study.

Factorial experiments are efficient at evaluating the effects and possible interactions of several factors (independent variables). Analysis of experiment design is built on the foundation of the analysis of variance, a collection of models that partition the observed variance into components, according to what factors the experiment must estimate or test. The split plot designs and Lattice designs are a very common subject for any forest statistician.

2.1 Split-plot Design

Split-plot designs are needed when the levels of some treatment factors are more difficult to change during the experiment than those of others. The designs have a nested blocking structure: split plots are nested within whole plots, which may be nested within blocks.

In simple terms, a split-plot experiment is a blocked experiment, where the blocks themselves serve as experimental units for a subset of the factors. Thus, there are two levels of experimental units. The blocks are referred to as whole plots, while the experimental units within blocks are called split plots, split units, or sub-plots. Corresponding to the two levels of experimental units are two levels of randomization. One randomization is conducted to determine the assignment of block-level treatments to whole plots. Then, as always in a blocked experiment, a randomization of treatments to split- plot experimental units occurs within each block or whole plot (Nachtsheim 2009).

2.2 Lattice Design

Yates (1936) developed an important type of incomplete block design named 'Lattice designs'. These designs were initially termed as pseudo-factor designs and were capable of providing efficient incomplete block designs for unstructured treatments by inflicting artificial pseudo-factor on the treatments and then confounding the contrasts between blocks. Lattice designs are now frequently used in the field of forestry.

The two most commonly used lattice designs include balanced lattice and partially balanced lattice. The discrepancy between these two designs occurs on the use of the number of replications. In balanced lattice design, block size (k) is equal to the square root of the total number of treatments and the number of replications required is one more than the block size i.e., $k + 1$. However, in partially balanced lattice design any number of replications can be used so with two replications the design is called a simple lattice; with three replications, a triple lattice; with four replications, a quadruple lattice, etc. (Raza and Masood 2009).

3

Diameter and Height Growth

Measurements of individual tree diameters and heights are commonly applied in most forest inventory situations to obtain estimates of growing stock. Height and diameter are two of the most important elements of tree size and forest structure. The development of the height and diameter of trees over a period of time in a stand not only characterizes stand development but also provides insight about the yield potential of the site. A crucial factor for sound management of any forest stand is the knowledge of the growth of individual trees and stands. A height-diameter model is an indispensable element of any individual tree-based growth-based projection system.

The h/d ratio is widely used in Europe in order to quantify the risk of storm damage and, in the thinning practice, to help with singletree classification (Oliveira 1987). According to Abetz, trees with lower h/d ratio have bigger volume increment and show a bigger increment increase after thinning (Kramer 1984). It has been observed that for single trees, value of h/d ratio of 60 to 80 is suitable for the selection of best ones.

1. DIAMETER

Diameter is a convenient variable for predicting the growth of trees in the plantations and forests. Being closely correlated with stem volume and value, diameter is easy to measure and is an essential quantity for economic and silvicultural decision-making. Furthermore, the diameter is easy to measure and initial diameter distributions are often available from forest inventories, providing a reliable base for projections. Quadratic mean diameter is an important element of a stand growth model. This quantity, which is essential for estimating product yield, may be derived from the per-hectare basal area and number of stems per ha. Modelling its growth usually proves to be easy, given data from permanent plots grown at different densities. However, problems usually arise when model is applied in thinned stands, where the mean diameter may develop in totally different ways, depending on the type of thinning (Gadow and Hui 1999, p 39).

The followings are some of the most frequently used modifications of the exponential function in forestry:

Schumacher (1939)

$$y = a\exp\left(\frac{-b}{x}\right) \qquad ...(3.1)$$

Gompertz (Medawar 1940, Yang et al 1978)

$$y = a \exp[-b \exp(-ex)] \quad ...(3.2)$$

Chapman-Richards (Richards 1959, Chapman 1961)

$$y = a[1 - \exp(-bx)]^c \quad ...(3.3)$$

where y is diameter and x represents age.

These growth curves may be fitted to diameter data to develop diameter-age functions. A serial correlation is often encountered in repeated measurements of the same plots. Hence, a 2-stage analysis is considered to be appropriate in such situation. For fitting the growth curves, first the data from the sample plots should be separated on the basis of site class, then within each site class the data may be grouped into different stand densities. After that the Schumacher, Gompertz and Chapman-Richards functions, given above, may be applied to these grouped data sets in each of the site classes and based on the fit statistics (R^2, RMSE and significance of parameter estimates), the best model may be selected for developing growth functions. The parameters of the selected function may further be modelled against the stand densities in each site class to generate equations. Further, parameter estimates of these equations may again be modelled against the different site classes to get the relationship between the parameters and the site classes. In this way, diameter growth functions may be developed. Following this process, diameter growth model has been developed for *Eucalyptus camaldulensis* (Tewari et al 2002) and *Dalbergia sissoo* (Tewari and Kumar 2005) planted in the Indira Gandhi Canal Project (IGNP) area of Rajasthan.

1.1 Diameter Distribution

The use of diameter distributions for predicting forest development has a special appeal and a rather long tradition in even-aged growth modelling (Meyer 1930, Prodan 1953, Kennel 1972, Bailey and Dell 1973, Hafley and Schreuder 1977). Unimodal diameter frequency distributions are often used to describe forest structure. Normal, lognormal, beta, gamma, Weibull and S_B are some of the statistical distribution functions that can be fitted to the diameter data. There is a great variety of empirical forest structures that can be described using a theoretical diameter distribution. Weibull distribution has been shown to be useful for describing diameter distributions in even-aged stands (Baily and Dell 1973, Schreuder and Swank 1974, Schreuder et al 1979). Hafley and Schreuder (1977) concluded that univariate S_B distribution fits the marginal diameter frequency consistently better than Weibull, beta, gamma, lognormal and normal distributions. However, this was not confirmed in case of Weibull (Gadow 1987). DBH frequencies may occur as bimodal distributions which represent more irregular forest structures and, hence, in such cases more flexible functions may be used. If the distribution is multimodal or largely skewed, a situation often encountered in the natural forests and in plantations under silvicultural management, it is assumed that Weibull and S_B distributions will likely to give better results than the normal distribution (Tewari 2000).

1.1.1 Weibull Distribution

The cumulative distribution function of X gives the probability that a value less than or equal to X will occur:

$$F(X) = P(x \leq X) = 1 - e^{-\left(\frac{X-a}{b}\right)^c} \quad ...(3.4)$$

with x = random diameter at breast height [DBH, cm]

X = DBH, for which the probability is calculated that x assumes a smaller value

$F(X) = P(x \leq X)$ = cumulative probability of the random variable x

= probability that a random DBH is smaller than X

a, b, c = location, scale and shape parameter of the Weibull distribution

The model has the facility for describing both positive and negative skew and approximating symmetry (Hafley and Schrueder 1977). One of the simplest and fastest methods of estimating Weibull parameters is the method of moments (Garcia 1981). When Weibull parameter '*c*' is equal to 3.6, the distribution will become normal distribution.

1.1.2 Univariate S_B distribution

The univariate Johnson S_B distribution density function (Johnson 1949a) is given by

$$f(x) = (\delta / \sqrt{2\pi}) \{\lambda / (x-\varepsilon)(\varepsilon+\lambda-x)\} \exp \{-1/2[\gamma+\delta \ln \{(x-\varepsilon)/(\varepsilon+\lambda-x)\}]^2\}$$

where,

$\lambda + \delta \ln \{(x-\varepsilon)/(\delta + \lambda - x)\} = Z$ (standard normal variate)

$\varepsilon < x < \varepsilon + \lambda;\ \delta > 0;\ -\infty < \gamma < +\infty;\ \lambda > 0;\ -\infty < \varepsilon < +\infty$

ε = minimum x, λ = range of x, γ and δ are shape parameters.

The tree diameter (D) is said to have a S_B distribution if

$$Z_D = \gamma_D + \delta_D \ln [(D - D_{min})]/(D_{max} - D) \quad ...(3.5)$$

has a standard normal distribution (mean=0, variance=1). Here, γ_D and δ_D are both scale and shape parameters and D_{min} and D_{max} are the lower and upper bounds for the variable D.

2. HEIGHT

Stand height is a key variable in most growth models. The height development of the size-class representative trees in an even-aged stand is rarely modelled directly. Height measurements are time-consuming and often inaccurate and initial height distributions are thus usually not available for projection. Instead, the heights are derived indirectly from the diameters, using a known or estimated relationship between diameters and heights (Van Laar and Akça 1997, p. 149) and modelling the development of this relationship over time. The relationship between diameters and heights may be described using a height regression or a bivariate diameter-height distribution. In practice, all the trees in the plots are measured for diameter and sub-sampling is done for height. The data from trees sampled for height are then used to develop a diameter-height regression, which in turn, is used to estimate height of other trees in the plot (Arabatizis and Burkhart 1992). At least 30 different functions have been used for describing the relationship between tree diameters and heights (Wenk et al 1990, p. 221). Some of the widely used models for height, diameter and age relationships are:

Schumacher (1939)

$$H = a \times e^{(b/t)} \quad ...(3.6)$$

Mirkovic (1958)

$$h_i = 1.3 + (a_1 + a_2 \times H_0 + a_3 \times D_g) \times e^{(-a_4/di)} \quad ...(3.7)$$

Prodan (1965)

$$D = t^2/(a + bt + ct^2) \quad ...(3.8)$$

Curtis (1967)

$$\ln H = a + b \times (1/D) \quad ...(3.9)$$

Gaffrey (1988)

$$h_i = 1.3 + (H_g - 1.3) \times e^{\{a_1 \times (1\text{-}Dg/di) + a_2 \times (1/Dg\text{-}1/di)\}} \quad ...(3.10)$$

Pienaar et al (1991)

$$h_i = a_1 \times h_0 \times [1 - a_2 \times e^{-\{a_3 \times (di/Dg)\}}]^{a_4} \quad ...(3.11)$$

Chapman-Richards (Clutter et al 1983)

$$H = a \times (1 - e^{kt})^{1/m} \text{ or, } H = a \times (1 - b \times e^{kt})^{1/(1\text{-m})} \quad ...(3.12)$$

where,

H = tree height (m)

D = diameter (cm), t = stand/tree age (yrs)

$D(t)$ = diameter at age t (cm)

h_i = height corresponding to the ith diameter class (m)

di = i[th] diameter class (cm)

H_0 = dominant stand height (m)

H_g = height of the av. basal area tree (m)

D_g = diameter of av. basal area tree (cm)

a, a_1, a_2, a_3, a_4, b, k and m are the parameters to be estimated

The equation (3.9) has wide acceptance in forestry applications (for example, Oliveira 1980, Schreuder and Hafley 1977, Gadow 1990, Tewari and Gadow 1999). In any tree stand, height growth of the trees will not show linear and constant growth rate with respect to diameter over a period of time and hence in the plots, height-diameter regression needs to be performed every time a re-measurement is done for volume estimation. To avoid this problem to some extent, the parameters 'a' and 'b' of equation (3.9) may be modelled against the stand age using the following parabolic equation:

$$X = a_1 + a_2 t + a_3 t^2 \quad ...(3.13)$$

where, t is the stand age and X is parameter 'a' or 'b' of equation (3.9).

Using this method, Tewari and Gadow (2003) developed height-diameter relationship for a pure even-aged *Prosopis cineraria* stands established in 1988 in the IGNP area of Jaiselmer district in Rajasthan (India) after laying out of a sample plot consisting of 154 trees in 1995 and taking measurements continuously for four years till 1998.

3. GENERALIZED DIAMETER-HEIGHT RELATIONS

A height regression may be derived separately for each stand, on the basis of pairs of diameter-height measurements obtained during a stand inventory. As this approach is costly, a practical alternative is to develop generalized height regressions which embody certain basic characteristics inherent in all individual height regressions (Wiedemann 1935, Prodan 1965, Kramer and Akça 1987, Wenk et al 1990, p. 223, Hui and Gadow 1993a, Prodan et al 1997, Hradetzky 1999).

Generalised diameter-height equations are useful tools for forest inventory purposes. Commonly, they are also used as important element of many size class models for simulating the development of silvicultural alternatives over time (Páscoa 1987). They are essential for estimating product yields for representative trees. Additionally, these are sometimes used to generate individual tree height increment data (Schröder and Gonzalez 2001).

The generalised diameter-height equations are different with the ordinary diameter-height equations in the sense that they include quadratic mean diameter, stand basal area or stems/ha as extra independent variable so that the equation can be applied on the plantations available on different sites with varying stocking. Only few such equations are available in the literature, which are modifications of Richards or Schumacher functions.

Equation may be derived on the basis of a function involving the logarithm of height as the dependent variable, and the reciprocal of diameter as the predictor variable.

$$\ln(Y) = a_0 + a_1 \cdot \frac{1}{X} \qquad \text{...(3.14)}$$

where, Y = some measure of tree dimension or stand production

X = predictor variable, e.g., stand age

a_0, a_1 = parameters

Numerous authors subsequently used this model, which is known as the *Schumacher function* or as the *Michailoff function*. This equation was modified and used by several authors as a generalized height regression:

$$h_i = 1.3 + (\bar{H} + 1.3)e^{a_1\left(1 - \frac{\bar{D}}{d_i}\right) + a_2\left(\frac{1}{\bar{D}} - \frac{1}{d_i}\right)} \qquad \text{...(3.15)}$$

with $\bar{H}$ = average stand height [m]

$\bar{D}$ = average stand diameter [cm]

h_i = height of i-th tree [m]

d_i = breast height diameter of i-th tree [cm]

a_1, a_2 = parameters to be estimated

An alternative model was developed by Hui and Gadow (1993a) on the basis of the allometric growth theory. However, instead of postulating a constant ratio of the relative growth rates of the two parts of the organism (Kramer 1988, p. 50), it was assumed that the ratio is influenced by a stand attribute such as the dominant height:

$$\frac{\frac{1}{h_i}\frac{\partial h_i}{\partial t}}{\frac{1}{d_i}\frac{\partial d_{i0}}{\partial t}} = aH^b \qquad ...(3.16)$$

where, t = time

H = dominant stand height [m]

h_i = height of i-th tree [m]

d_{i0} = diameter at the base of the i-th tree [cm]

a, b = parameters

Integration of this equation gives:

$$\ln h_i = aH^b \ln d_{i0} + k \qquad ...(3.17)$$

with, k = constant of integration

This may be further developed so that eventually a generalized height regression is obtained.

$$h_i = 1.3 + a_1 H^{b_1} d_i^{a_2 H^{b_2}} \qquad ...(3.18)$$

with, H = dominant stand height [m]

h_i = height of i-th tree [m]

d_i = breast height diameter of i-th tree [cm]

a_1, a_2, b_1, b_2 = parameters to be estimated

This equation proved to be very suitable when compared with other models, a particular advantage being its simplicity. Dominant height, representing the stand level, is the only additional variable required for estimating individual tree heights from diameters.

Pienaar (1991) derived the following equation from the Richards function (cp. Richards 1959) for stands of slash pine in the south-eastern United States:

$$h_i = \alpha_1 H_0 (1 - \alpha_0 e^{-\beta d_i / D_g})^{\alpha_3} \qquad ...(3.19)$$

Mirkovich (1958) derived the following equation from the Schumacher function (cp. Schumacher 1939) for oak stands in central Europe (see also Michailoff 1943, Prodan et al 1997):

$$h_i = 1.3 + (\alpha_i + \alpha_2 H_0 - \alpha_3 D_g) e^{-\beta / d_i} \qquad ...(3.20)$$

It can be easily shown that stands with the same D_g can have different stand densities and hence it is interesting to see whether model can be further improved by incorporating a variable, which accounts for stand density. Thus, Schröder and González (2001) tried to improve and modify equation (3.20) as follows by incorporating stand basal area as an independent variable:

$$h_i = 1.3 + (\alpha_i + \alpha_2 H_0 - \alpha_3 D_g + \alpha_4 G)e^{-\beta/\sqrt{d_i}} \quad ...(3.21)$$

In equations (3.19 to 3.21), the symbols used are:

h_i = height of tree i (m)

d_i = breast height diameter over bark of tree i (cm)

H_0 = dominant stand height (m)

D_g = quadratic mean diameter of the stand (cm)

G = stand basal area (m^2/ha)

α_1, α_2, α_3, α_4, β_1, β_2=parameters to be estimated

1.3 = a constant used to avoid the prediction of a height <1.3 m when d_i is very small

Tewari (2006) used the above equations to develop generalised diameter-height equations for pure stands of neem (*Azadirachta indica*) in semi-arid region of Gujarat State in India and found that equation (3-18) performed best with parameter estimates a_1 = 0.17856, a_2 = 1.62130, b_1 = 1.09787, b_2 = -0.52973. The generalised height-diameter equation was also developed for the *Tecomella undulata* plantations established by the State Forest Department of Rajasthan in the Indira Gandhi Canal Project (IGNP) area which is under drought prone hot arid region (Tewari 2007a) and again equation (3-18), the model proposed by Hui and Gadow (1993a), performed best among all the four equations tested with the parameter estimates estimates a_1 = 0.412, a_2 = 0.571, b_1 = 0.436, b_2 = 0.064.

4. BIVARIATE DIAMETER-HEIGHT DISTRIBUTIONS

A height-diameter regression is often not very appropriate because of great variations of height for a given diameter. It is thus more realistic to assume a distribution of heights corresponding to a given diameter class. Information about size-class distributions of the trees within a forest stand is important for estimating product yields. The size-class distribution influences the growth potential and hence the current and future economic value of a forest (Knoebel and Burkhart 1991).

An important problem in forestry is the prediction of stand yields in volume on the basis of stand age, productive capacity of the site and stand density. Forest managers are more interested in studying the management practices like thinning, fertilizations on volume in pulp wood, sawn timber and veneer. The volume in these products is heavily dependent on tree diameter and height distribution. Hence there is considerable interest is successful fitting of a bivariate statistical distribution to describe diameter-height frequency data.

The bivariate extension of the univariate S_B distribution (Johnson 1949a), which is known as the S_{BB} distribution (Johnson 1949b), may provide useful information by describing even-aged forest height-diameter data (Hafley and Schreuder 1977).

The S_{BB} is a *bivariate* distribution for which both marginals[1] are S_B distributions. It is the bivariate distribution of y_1 and y_2 when the standard normal variates z_1 and z_2 are defined as:

$$z_1 = \gamma_1 + \delta_1 \ln \{y_1/(1-y_1)\} \text{ and } z_2 = \gamma_2 + \delta_2 \ln \{y_2/(1-y_2)\} \quad ...(3.22)$$

where z_1 and z_2 have the joint bivariate distribution with correlation ρ, namely,

$$p(z_1, z_2; \rho) = [2\pi\sqrt{1-\rho^2}\,]^{-1} \exp\{-(1/2)(1-\rho^2)^{-1}(z_1^2 - 2\rho z_1 z_2 + z_2^2)\} \quad ...(3.23)$$

Here, $y_1 = (d-\xi_1)/\lambda_1$ and $y_2 = (h-\xi_2)/\lambda_2$ and ξ_1 and ξ_2 represent the smallest values, and λ_1 and λ_2 the range of diameters (d) and heights (h), respectively, in the population. Many of the properties of the S_{BB} can be obtained directly from the literature of the bivariate normal distribution, because of the relation between them.

The correlation coefficient is obtained from

$$\rho = \sum_{j=1}^{n} z_{1j} z_{2j}/n \quad ...(3.24)$$

where, $z_{ij} = \gamma_i + \delta_i \ln \{y_{ij}/(1-y_{ij})\}$; $y_{ij} = (X_{ij} - \xi_i)/\lambda_i$ with $X_{1j} = d_j$ and $X_{2j} = h_j$; n = number of diameter-height pairs of data; $i = 1, 2$; $j = 1, n$. The parameters and are estimated as follows:

$$\gamma_i = (-f_i)/s_i \quad ...(3.25)$$

$$\delta_i = 1/s_i \quad ...(3.26)$$

where, $f_i = \left(\sum_{j=1}^{n} f_{ij}\right)/n$; $s_i^2 = 1/n \sum_{j=1}^{n} (f_{ij} - f_i)^2$; $f_{ij} = \ln\{y_{ij}/(1-y_{ij})\}$; n and y_{ij} is as above; $i = 1, 2$; $j = 1,..., n$

One of the properties of interest is the regression relation between y_2 and y_1. In general, the usual mean regression is complicated. However, the median regression takes a much simpler form, namely:

$$y_2 = \theta y_1^{\phi} \{(1-y_1)^{\phi} + \theta y_1^{\phi}\}^{-1} \quad ...(3.27)$$

where,

$$\theta = \exp \{(\rho\gamma_1 - \gamma_2)/\delta_2\} \text{ and } \phi = \rho\delta_1/\delta_2$$

Here, $\phi > 0$ since it is assumed that $\rho > 0$. The shape of the regression curve is influenced by ϕ, and the slope depends on the range of values of ρ. For $0 < \phi < 1$, the first derivative of y_2 with respect to y_1 does not exist at the extremes, $y_1 = 0$ and $y_1 = 1$. For

1. Relative to the joint density $f_{X,Y}(x,y)$, the densities of X and Y, $f_X(x)$ and $f_Y(y)$, are called the marginal densities. From a practical perspective, the marginal distributions describe the individual behavior of the measurements X and Y, but they often lack, even when taken together, information regarding the joint behavior of the measurements. Although the marginal densities can be recovered from the joint density, in general, the joint density can not be constructed from a knowledge of the marginal densities. In the discrete case, the marginal density $f_X(x)$ is found by holding x fixed and summing over y, whereas in the continuous case, it is found by holding x fixed and integrating over y. An analogous comment applies to finding $f_Y(y)$ (see Dougherty 1990, p. 208-9 & 215).

data sets of diameter and height, the constraint $\phi > 1$ should be imposed on parameter estimation. The regression is linear if $\rho\delta_1 = \delta_2$ and $\rho\gamma_1 = \gamma_2$. An important aspect of the equations is that and are functions of the two marginal distribution parameters. Rewritten in terms of diameter and height, Equation (3.27) can be expressed as

$$(h - \xi_2)/\lambda_2 = \theta[\{(\xi_1 + \lambda_1 - 1)/(d - \xi_1)\}^{\phi} + \theta]^{-1} \quad ...(3.28)$$

A second property of interest is the fact that the conditional distribution of y_2 given y_1 also follows the S_B distribution with parameters 'γ and δ'.

where,

$$\gamma' = [\gamma_2 - \rho z_1](1 - \rho^2)^{-1/2} \text{ and } \delta' = \delta_2(1 - \rho^2)^{-1/2}.$$

This conditional distribution can be bimodal[2].

Tewari and Gadow (1997) fitted the distribution to the diameters and heights of a 10-year-old *Acacia tortilis* plantation consisting of 251 trees, located in the hot arid tract of Rajasthan (India). Diameters were measured to the nearest 0.1 cm and heights to the nearest 5 cm. The statistics for the diameter-height data of *Acacia tortilis* are given in Table 3.1.

Table 3.1: Distribution parameters for diameters and heights of *Acacia tortilis*.

Distribution parameters	**Diameters (cm)**	**Heights (m)**
Mean	7.11	4.47
Variance	4.78	0.55
Index of Skewness	0.3207	-0.3263
Index of Kurtosis	3.0680	3.1016

The diameter-height data pairs were used to fit the S_{BB} distribution using the computer programme 'GETIT.PAS (FIT S_{BB})' developed by W.L. Hafley. Firstly, the data were used to estimate the parameters of marginal S_B distribution. The results are presented in Table 3.2.

---

2. The extraction of information about one random variable, when given information about the behavior of another, leads to the problem of conditioning. If *X* and *Y* are discrete and posess the joint density F(x, y), then conditional probabilities regarding *Y* take the form of conditional probabilities concerning the relevant events. The probability that $Y = y$, given that $X = x$, is

   $P(Y = y \mid X = x) = P(Y = y, X = x)/P(X = x) = f(x, y)/f_x(x)$

   Intuitively, the observation $X = x$ is recorded, and one is left with probabilistic knowledge concerning the random variable *Y*. Since *Y* is discrete, only the conditional probabilities of its possible outcomes need to be specified. In the case of jointly distributed continuous random variables, as long as $f_x(x) \neq 0$, $f(x, y)/f_x(x)$ is still defined. This expression is taken as the general definition of the conditional density. The random variable associated with the conditional density is known as the conditional random variable of *Y* given *x*, and is denoted by $Y|x$ (see Dougherty 1990, p. 249).

Table 3.2: Parameters of the S_B distributions for diameters and heights of *Acacia tortilis.*

SB parameters	**Diameter**	**Height**
Gamma (γ)	0.9686	-0.5729
Delta (δ)	1.8809	1.6460
Mu (μ)	-0.5150	0.3480
Sigma (σ)	0.5317	0.6075

The parameters γ_i and δ_i (i = 1, 2) were used to transform the observation pairs to standard normal deviates which, in turn, were used to calculate the marginal S_B densities[3] (Table 3.3).

Table 3.3: Marginal S_B distributions separately for diameters and heights of *Acacia tortilis.*

Diameter				**Height**			
Class bounds	**Observed frequency**	**Predicted frequency**	**Residual**	**Class bounds**	**Observed frequency**	**Predicted frequency**	**Residual**
0.0 - 2.0	2.0	0.3	-1.7	1.3 - 2.0	0.0	0.0	-0.0
2.0 - 4.0	15.0	17.4	2.4	2.0 - 4.0	76.0	68.1	-7.9
4.0 - 6.0	64.0	65.8	1.8	4.0 - 6.0	173.0	181.0	8.0
6.0 - 8.0	90.0	84.0	-6.0	6.0 - 6.8	2.0	1.9	-0.1
8.0 - 10.0	59.0	56.7	-2.3				
10.0 - 12.0	16.0	22.1	6.1				
12.0 - 14.0	5.0	4.4	-0.6				
14.0 - 18.6	0.0	0.3	-0.3				

The next step was to calculate the correlation coefficient using the standard normal deviates in equation (3.24). Further, the correlation coefficient ρ and marginal S_B distribution parameters were substituted in equation (3.23) to obtain the S_{BB} parameters (Table 3.4).

Table 3.4: Parameters of the bivariate S_{BB} distribution.

S_{BB} parameters	**Values**
Rho (ρ)	0.5428
Theta (θ)	1.9492
Phi (ϕ)	0.6200

Finally, the correlation coefficient and the standard normal deviates z_{ij} were used to obtain the S_{BB} joint density distribution (Table 3.5).

3. We need to point out that in this case the marginal densities are not recovered from the bivariate S_{BB} joint distribution but obtained by fitting the univariate S_B distribution model.

Table 3.5: Observed and predicted S_{BB} joint distributions for the height and diameter classes.

Diameter class (cm)	Height class (m)					
	2.0 - 4.0		**4.0 - 6.0**		**6.0 - 6.8**	
	obs.	**pred.**	**obs.**	**pred.**	**obs.**	**pred.**
0.0 - 2.0	0.0080	0.0011	0.0000	0.0001	0.0000	0.0000
2.0 - 4.0	0.0438	0.0475	0.0159	0.0218	0.0000	0.0000
4.0 - 6.0	0.1155	0.1144	0.1394	0.1478	0.0000	0.0000
6.0 - 8.0	0.0996	0.0793	0.2590	0.2547	0.0000	0.0007
8.0 - 10.0	0.0319	0.0249	0.1992	0.1988	0.0040	0.0021
10.0 - 12.0	0.0000	0.0037	0.0598	0.0814	0.0040	0.0028
12.0 - 14.0	0.0040	0.0002	0.0159	0.0156	0.0000	0.0016

The primary advantage of the S_{BB} approach is that it allows for the generation of bivariate frequencies for height and diameter, thus being a more realistic model than the traditional regression approach.

Tewari and Gadow (1999) modelled the relationship between tree diameters and heights using S_{BB} bivariate distribution for even aged plantation of *A. tortilis* and found that the S_{BB} regression produced a better fit compared to usual mean regression. The S_{BB} median regression generally provides better results if the condition $\phi > 1$ is strictly followed. Sometimes, an estimated $\phi < 1$ does not support the assumption of a S_{BB} distribution using standard fitting procedures. To fulfil this condition, one may increment the estimate of the diameter range parameter by the width of a diameter class until $\phi > 1$ is attained. The bivariate S_{BB} distribution is more realistic and provides more useful information than the currently accepted mean regression approach in describing forest stand height-diameter data.

4

Tree Volume and Stand Yield

1. VOLUME EQUATIONS

One of the most important operations in forest mensuration is estimating the volume of sample trees (Robinson and Wood 1994). Estimates of volume are an important part of a stand model. Such estimates are indispensable when silvicultural decisions are based on economic criteria. Accurate estimates of tree volume are fundamental for forest ecosystem modelling and regional carbon accounting. Volume equations are critical starting points if forest management is to be successful and efficient. Allometric equations for predicting wood volume play a critical and obvious role in the management of any silvicultural system, and their absence would represent an impediment to developing and implementing management plans geared towards the harvest and utilization of wood products.

Estimation of tree volume with greater accuracy has always been a matter of interest for forest managers. Construction of volume equations for tree species is an important step in this direction. The importance of volume equations is well indicated by the existence of numerous such equations and the constant search for their improvement.

It is impossible to measure the stand volume directly in the field. Therefore, it is calculated from other variables, such as basal area and dominant height, and sometimes from stand age. During the past few decades, forest scientists have increasingly come to utilize regression for the purpose of constructing prediction equations. Some of the volume equations currently in practice are given below in the Table 4.1.

Table 4.1: Some important volume equations frequently use in forestry literature

Model	Independent variables
Combined variable	D^2H
Naslund	D^2, D^2H, H^2, DH^2
Comprehensive	D, DH, D^2, H^2, D^2H
Australian	D, H, D^2H
Meyer	D, DH, D^2, D^2H
Constant form factor	D^2H

In destructive sampling, the stem of the tree is divided into logs of definite sizes (say 3m) and then log volume is estimated using different formulas. Some of them are given in the Table 4.2.

Table 4.2: Formulas for calculating log volume

Name	Formula
Newton	V = [(B + 4M + S)/6]*L
Huber	V = M*L
Smalian	V = [(B + S)/2]*L
Bruce	V = (0.25*B + 0.75*S)*L
Centroid	$V = S*L + bL^2/2 + cL^3/3$
Sorenson	$V = f*(ds + 0.05*L)^2*L$
Subneiloid	$V = f*[(ds + dl)/2]^2*L$
Two-end conic	$V = f*[(ds^2 + ds*dl + dl^2)/3]*L$

where,

$b = (B\text{-}S\text{-}cL^2)/L$

$c = [B\text{-}C(L/e)\text{-}S(1\text{-}L/e)]/(L^2\text{-}Le)$

C = cross sectional area at mid volume of log measured at a distance q from the butt end of the log where, $q=L((((((D/d)^4+1)^{0.5}-2^{0.5})/(2^{0.5}((D/d)^2-1))))L)$

e = L-q

D, d = diameter (cm) at butt and small end respectively of the log

L = log length (m)

B = cross sectional area at butt end of log (m^2)

M = cross sectional area at mid length of log (m^2)

S = cross sectional area at small end of log (m^2)

f = 0.00007854

ds, dm, dl = small, mid-length and large-end diameters (cm)

Estimation of butt-log volumes has long been a problem for field foresters. It is well recognized that a measurement of diameter at the mid-length of a log, in addition to diameter measurements at each end, will improve the volume estimates (Wiant et al 1992, Patterson et al 1993). Such measurements are difficult to take with stacked logs. The measurement of mid-length diameter required to apply Huber's formula may not be taken on stacked logs.

Tewari and Singh (2005) used a database consisting of measurements of 167 logs collected from the sample trees of *Eucalyptus hybrid* (73 logs) and *Acacia nilotica* (94 logs) from Gujarat State (India) to compare the performance of the Huber, Smalian, Newton, Sorenson, Bruce, subneiloid and two-end conic methods for estimating volumes of butt logs and found that the Bruce formula performed better in comparison to other models as far as composite effect of accuracy and precision is concerned.

To construct volume equation through destructive sampling, trees of different diameter classes may be felled and their total height (H), diameter at breast height (D) and volume (V) should recorded. Trees representing different diameter classes may be felled on proportionate basis to have representative sample of the stand. The length of the felled tree may be measured with a tape and stump height should be added to get the total height. For the computation of total volume, stem and branch wood with a minimum diameter of 5 cm can be considered. The volume may then be calculated by dividing the stem and branches into logs of 3m length, measuring the mid-diameters and applying Huber's formula to estimate individual log volumes. Thereafter, All the log volumes may be added to arrive at total tree volume.

The role of validation in examining the predictive ability of a volume equation before its application in the field has been stressed by various authors (Goulding 1979, Reynolds et al 1981). For validation purpose, models can be tested in various ways. Firstly, the re-sampling approach (Bi and Hamilton 1998, Bi 1999, Bi 2000) may be suggested. Secondly, iterative validation procedure (Williams 1997) may be considered. Thirdly, if one has two independent data sets from the same area or population, one set can be used for fitting and the other for validation. Aalternatively, if one has only one data set, it can be divided into two through random procedure for the purpose and the first data set containing, e.g., 70% of the observations may be used for fitting the volume equations while the latter containing the remaining data can be used for validation.

The aim of any volume equation is to provide accurate estimates with acceptable levels of local bias over the entire range of tree size in the data. Equations that provide accurate predictions of volume without local bias over the entire range of diameter are one of the basic building blocks of a forest growth and yield simulation system (Bi and Hamilton 1998). The accurate prediction of intermediate and final harvests in the construction of yield tables depends on the accuracy of individual-tree volume equations (Perez and Kanninen 2003, Bi and Hamilton 1998).

Since growing rates, management regimens, and production objectives are specific to each country or world region, volume equations should not be generalized assuming that trees have similar bole shapes. For instance, different thinning and pruning regimens alter the stem form significantly (taper, wood defects, eccentricity, and height/dbh ratio) and consequently volume predictions may not be accurate if calculated independent of the stand characteristics (Perez and Kanninen 2003). Furthermore, commercial volume of important timber tree stands has been arbitrarily estimated from the visual classification of the trees in terms of number of logs up to a "commercial height" and a general diameter reduction factor, which shows the urgent need for merchantable volume equations.

For variable-bole length (top height) merchantable volume equations, equation (4.1) may be used. Merchantable volume equations of this type have been used by Kozak et al (1969), Avery and Burkhart (1994) and Perez and Kanninen (2003).

$$V_m = \pi * D^2 * (a*h + b*(h^2/H) + c*(h^3/H^2)) \quad ...(4.1)$$

Similarly, equation (4.2) may be used to develop variable-top-diameter merchantable volume equations. Similar equations have been used by many workers in the past (Honer 1964, Burkhart 1977, Clutter 1980, Perez and Kanninen 2003).

$$V_m = V * (1 - a * D_m^{b} * D^{c}) \quad ...(4.2)$$

where,

V_m = merchantable stem volume (m^3) to a top diameter D_m

D = diameter at breast height (cm)

h = upper stem merchantability limit (expressed as merchantable height in m)

H = total tree height (m) by direct measurement or estimated by a volume equation

V = total stem volume (m^3)

D_m = upper stem merchantability limit (expressed as minimum diameter in cm)

a, b, c = regression coefficients

Numerous volume equations have been developed, even for various tree species in India, which are available in literature. For example, total and merchantable wood volume equations have been developed for neem (Jain et al 1998), *A. nilotica* (Tewari and Singh 2006a), teak (Tewari and Singh 2018) and *E.* hybrid (Tewari and Singh 2006b; Fig. 4.1) plantations in Gujarat and forest teak plantations (Tewari et al 2013) in Karnataka. Volume equations have also been developed for irrigated plantations of *Acacia tortilis* (Jain et al 1996), *Dalbergia sissoo* (Tewari and Kumar 2001), *Eucalyptus camaldulensis* (Tewari and Kumar 2003) and *Tecomella undulata* (Tewari 2007b) in the Indira Gandhi Nahar Pariyojana (IGNP) area of Rajasthan. The volume equations (single & double-entry) developed for these species are given below:

E. camaldulensis (Tewari and Kumar 2003)

$$V = 0.000169*D^{2.41298}$$

$$V = -0.00226 + 0.0000333\ D^2H$$

D. sissoo (Tewari and Kumar 2001)

$$V = 0.01328 - 0.00538\ D + 0.000760\ D^2$$

$$V = -0.0023 + 0.0000364\ D^2H$$

T. undulata (Tewari 2007b)

$$V = 0.000088\ D^{2.381398}$$

$$V = 0.000066\ D^{2.100121}H^{0.553696}$$

A. tortilis (Jain et al 1996)

$$V = -0.02179 + 0.00003451*D^2H + 0.0001990*DH$$

$$\sqrt{(V)} = -0.09415 + 0.0253594*D$$

A. indica (Jain et al 1998)

$$V = 0.07033 - 0.01387*D + 0.0009853*D^2$$

$$V = -0.00989 + 0.0000184*DH + 0.0000438*D^2H$$

A. nilotica (Tewari and Singh 2006a)

$$V = 0.000071*D^{2.735778}$$

$$V = 0.000018*D^{2.363677}H^{0.938962}$$

E. hybrid (Tewari and Singh 2006b)

$$V = 0.000076*D^{2.761477}$$

$$V = 0.000014*D^{2.141947}H^{1.168588}$$

Tectona grandis (Tewari et al 2013)

$$V = 0.000128*D^{2.522838}$$

$$V = 0.000046*D^{2.136615}H^{0.785557}$$

Tectona grandis (Tewari and Singh 2018)

$$V = 0.000055*D^{2.775723}$$

$$V = -0.001109 + 0.000038\ D^2H$$

Variable bole-length volume equations

A. nilotica (Tewari and Singh 2006a)

$$V_m = \pi*D^2*(0.000040*h - 0.000103*(h^2/H) + 0.000068*(h^3/H^2))$$

E. hybrid (Tewari and Singh 2006b)

$$V_m = \pi*D^2*(0.000103*h - 0.000077*(h^2/H) + 0.000016*(h^3/H^2))$$

T. grandis (Tewari et al 2013)

$$V_m = \pi*D^2*(0.000028*h - 0.000024*(h^2/H) + 0.000008*(h^3/H^2))$$

Variable top-diameter volume equation

T. grandis (Tewari et al 2013)

$$V_m = V * (1 - 0.432896 *Dm^{3.064809}*D^{-2.868943})$$

where, V is the total wood volumes (m^3) over-bark, V_m is merchantable stem volume (m^3) over-bark to a top height h, D is the dbh (cm) and H is the total tree height (m).

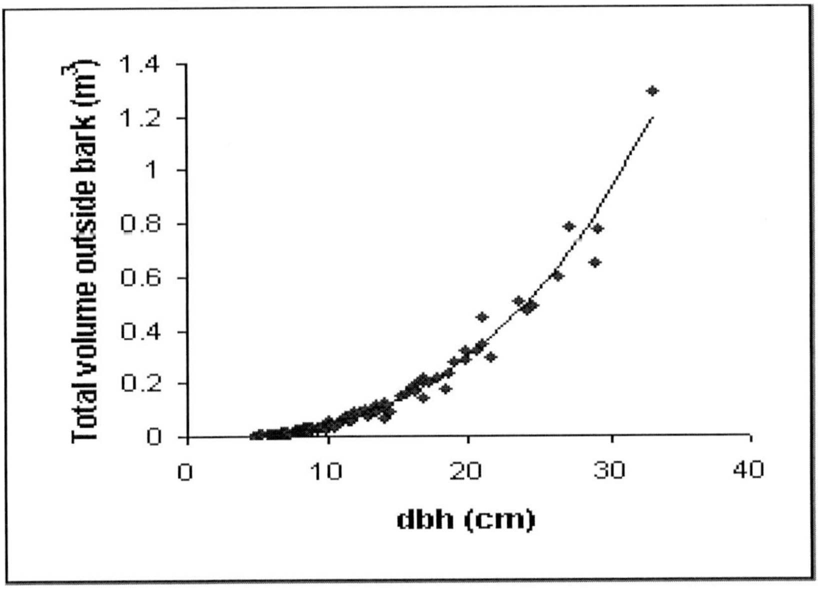

Fig. 4.1. *Volume vs. dbh curve for E. hybrid plantations in Gujarat*

1.1 Taper Functions

Stem profile equations (Taper functions) help in Estimating tree volume with greater accuracy. A number of regression models have been developed to describe stem profile, using total height as one of the variables (Bruce et al 1968, Kozak et al 1969, Demaerschalk 1972, Max and Burkhart 1976, Demaerschalk and Kozak 1977, Brink and Gadow 1986, Riemer et al 1995). These models predict upper stem diameters at any height and most of them can be transformed to estimate the volume of a given stem section.

Many of the hardwood species show a decurrent branching pattern. Because of indistinct terminals and visual interference by tree crowns, it is difficult to measure total height of the tree with decurrent form (Fowler and Rennie 1988, Philip 1994, Kramer and Akça 1995). In these species, the height up to the base of the live crown (referred to as merchantable height, MHT) can easily be measured. Therefore, it is pertinent to develop taper equations for these species based on merchantable height instead of total height. This is desirable particularly in the case where high-quality sawn timber is a primary concern.

Different taper equations have also been developed to estimate the tree/log volume. Taper functions, mathematical models for describing the shape and taper of tree boles, are effective for estimating following three important attributes of a tree which are important in partitioning a tree into a variety of products:

1. Diameter at any point along the bole
2. Height to any predetermined diameter
3. Volume between any two points on the bole

Diameter at the crown base is subtracted from all measured stem diameters including dbh of each tree so that the transformed diameter at crown base is zero which provides compatibility with stem profile equations while maintaining tree stem profile. The over-bark volumes are calculated by summing the volumes of each stem section. The volume of the first section, from 0.0 to 0.10 m, may be calculated by assuming the shape of a cylinder while the volume of the last section may be calculated by assuming the shape of a cone (Byrne and Reed 1986). The volumes of the remaining sections can be calculated using the Smalian formula (Avery and Burkhart 1994).

The following are some of the stem profile models based on total tree height used to fit to the transformed data as explained above:

Kozak's Function (Kozak et al 1969)

$$\frac{d'^2}{D'^2} = \alpha\left(1 - \frac{2h}{MHT} + \frac{h^2}{MHT^2}\right) \qquad ...(4.3)$$

Demaerschalk's Function (Demaerschalk 1973)

$$\frac{d'^2}{D'^2} = \alpha\left(\frac{MHT\text{-}h}{MHT}\right)^{\beta} \qquad ...(4.4)$$

Modified Brink Function (Riemer et al 1995)

$$d' = i + (D' - i)\frac{e^{p(1.3-h)} - e^{p(1.3-MHT)}}{1 - e^{p(1.3-MHT)}} - i\frac{e^{q(h-MHT)} - e^{q(1.3-MHT)}}{1 - e^{q(1.3-MHT)}}$$

$$= u + ve^{-ih} - we^{qh} \qquad ...(4.5)$$

where,

$$u = \frac{i}{1 - e^{q(1.3-MHT)}} + (D - i)\left(1 - \frac{1}{1 - e^{p(1.3-MHT)}}\right);$$

$$v = \frac{(D' - i)e^{1.3p}}{1 - e^{p(1.3-MHT)}} \text{ and } w = \frac{ie^{-q\,MHT}}{1 - e^{q(1.3-MHT)}}$$

The notations used in the functions mentioned above are as under:

d' = transformed diameter (diameter outside bark at any given h minus diameter outside bark at MHT, cm)

h = height above the ground (m)

MHT = merchantable height (m)

D′ = transformed dbh (dbh outside bark minus diameter outside bark at MHT, cm)

α and β = regression coefficients

i = Parameter ('common Asymptote' Brink and Gadow 1986)

p = Parameter (for below stem portion)

q = Parameter (for upper stem portion)

Model (4.3), having no intercept, was derived from a parabolic function. Model (4.4) was obtained from the volume over basal area as a function of height (without intercept) while model (4.5) resulted from the solution of a decay function. Model (4.5) complies with the conditions that observed dbh is exactly equal to model dbh and model diameter is exactly zero at crown base. The important constraint that observed dbh is exactly equal to model dbh is not imposed by models (4.3) and (4.4).

Using height up to the crown base in lieu of total height, Trincado et al (1996) applied the above three taper functions to develop stem profile equations for *Quercus robur* and concluded that model (4.5) provided better fit for predicting stem diameter followed by models (4.4) and (4.3).

The stem profile functions can be rearranged to predict height for a specified diameter and to calculate the volume between two specified heights. The resulting equations for the three models (4.3 to 4.5) assume the form as follows:

Kozak's Function (Kozak et al 1969)

$$h = MHT\left[1 - \frac{1}{\sqrt{\alpha}}\frac{d'}{D'}\right] \qquad ...(4.6)$$

$$v'_{(h_2,h_1)} = \frac{\pi D^2 \alpha}{4}\left[(h_2 - h_1) - \frac{1}{MHT}(h_2^2 - h_1^2) + \frac{1}{3MHT^2}(h_2^3 - h_1^3)\right] \qquad ...(4.7)$$

Demaerschalk's Function (Demaerschalk 1973)

$$h = \text{MHT} - \text{MHT}\left(\frac{d'^2}{\alpha \times D'^2}\right)^{\frac{1}{\beta}} \quad ...(4.8)$$

$$v'_{(h_2,h_1)} = -\frac{\pi D'^2 \alpha}{4(\beta+1)\text{MHT}^{\beta}}[(\text{MHT} - h_2)^{\beta+1} - (\text{MHT} - h_1)^{\beta+1}] \quad ...(4.9)$$

Modified Brink Function (Riemer et al 1995)

For this model, there is no closed solution for estimating tree height. The height must be determined iteratively, for a given upper stem diameter.

$$v'_{(h_2,h_1)} = \frac{\pi}{4}\left[u^2(h_2 - h_1) - 0.5\left\{\frac{v^2}{p}(e^{-2ph_2} - 2^{-ph_1}) - \frac{w^2}{q}(e^{2qh_2} - e^{2qh_1})\right\}\right.$$

$$\left.-2\left\{\frac{uv}{p}(e^{-ph_2} - e^{-ph_1}) + \frac{uw}{q}(e^{qh_2} - e^{qh_1})\frac{vw}{p-q}(e^{(q-p)h_2} - e^{(q-p)h_1}\right\}\right] \quad ...(4.10)$$

where,

$v'_{(h_2,h_1)}$ = Transformed volume between heights h_2 and h_1.

For estimating the real volume, transformed diameters d' should be replaced by real diameters d = d'+d_{top} in the equations (4.3), (4.4) and (4.5) for substitution in the integral $v_{(h_2,h_1)} = \frac{\pi}{4}\int_{h_1}^{h_2} d^2 dh$. The equations (4.7), (4.9) and (4.10) will assume the following form, respectively:

$$v_{(h_2,h_1)} = \frac{\pi D'^2 \alpha}{4}\left[(h_2 - h_1) - \frac{1}{\text{MHT}}(h_2^2 - h_1^2) + \frac{1}{3\text{MHT}^2}(h_2^3 - h_1^3)\right]$$

$$+\frac{\pi D\sqrt{\alpha}\, d_{top}}{4}\left[2(h_2 - h_1) - \frac{1}{\text{MHT}}(h_2^2 - h_1^2)\right] + \frac{\pi d_{top}^2}{4}(h_2 - h_1) \quad ...(4.11)$$

$$v_{(h_2,h_1)} = \frac{\pi D^2 \alpha}{4(\beta+1)\text{MHT}^{\beta}}[(\text{MHT} - h_2)^{\beta+1} - (\text{MHT} - h_1)^{\beta+1}]$$

$$-\frac{\pi D'\sqrt{\alpha}\, d_{top}}{2\left(\frac{\beta}{2}+1\right)\text{MHT}^{\frac{\beta}{2}}}\left[(\text{MHT} - h_2)^{\frac{\beta}{2}+1} - (\text{MHT} - h_1)^{\frac{\beta}{2}+1}\right] + \frac{\pi d_{top}^2}{4}(h_2 - h_1) \quad ...(4.12)$$

$$v_{(h_2,h_1)} = \frac{\pi}{4}\left[u^2(h_2 - h_1) - 0.5\left\{\frac{v^2}{p}(e^{-2ph_2} - e^{-2ph_1}) - \frac{w^2}{q}(e^{2qh_2} - e^{2qh_1})\right\}\right.$$

$$-2\left\{\frac{uv}{p}(2^{-ph_2}-e^{-ph_1})+\frac{uw}{q}(e^{qh_2}-e^{qh_1})-\frac{vw}{p-q}(e^{(q-p)h_2}-e^{(q-p)h_1}\right\}\Bigg]$$

$$+\frac{\pi d_{top}}{2}\left[\left(u+\frac{d_{top}}{2}\right)(h_2-h_1)-\frac{v}{p}(e^{ph_2}-e^{-ph_1})-\frac{w}{q}(e^{qh_2}-e^{qh_1})\right] \quad \text{...(4.13)}$$

Sometimes it may not be feasible to measure the top diameter. In such cases, a regression equation can be developed to estimate top diameter (d_{top}) from dbh using some sample trees which may be substituted in the equations discussed above. Eventually, only two variables, dbh and height, will be required.

Taper function systems have also been developed using importance sampling, control variate sampling and centroid sampling. Importance sampling, introduced by Gregoire et al (1986), gives unbiased estimates of bole volume based on one or two measurements of diameter. The system is quick, efficient and flexible. Control variate estimation was presented and compared with importance sampling by Gregoire et al (1987). It was also presented independently and concurrently by Van Deusen (1987) and Van Deusen and Lynch (1987) as a development of other system, the critical height method (Kitamura 1964). Control variate uses proxy bole to model the tree. Wood et al (1990) introduced centroid sampling, an adoption of importance sampling. It has two advantages over importance sampling. Since measurement is approximately at the centroid of the bole, dendrometry is not required on the tree in the upper crown where foliage often obscures the field of view. Prediction of volume is also more consistent from tree to tree.

1.2 Quarter-Girth Formula

The most widespread cubic log rule that includes an assumption regarding processing loss is quarter-girth formula, known as Hoppus rule. It was derived in Britain and is widely used internationally. The formula is:

$$V = (C/4)^2 * L/10{,}000 \quad \text{...(4.14)}$$

where,

V = log volume in m^3

C = log circumference in cm

L = log length in m

Quarter-girth formula gives 78.5% of the actual cubic volume of the log, making 21.5% lumber processing allowance for slabs, edgings, and sawdust. The formula can be converted to an estimating the full cubic volume of the log by multiplying by 1/0.785=1.2739. Multiplying the formula by adjustment results in:

$$V = (C/4)^2 * L/7{,}850 \quad \text{...(4.15)}$$

2. YIELD TABLES

A yield table is essentially a tool of long-term planning. It is a type of growth or 'experience' table which lists expected productivity/volumetric yield for a given age, site or crop quality and sometimes other indices such as density. Thus, yield tables usually refer only to even-aged stands. Data to prepare such tables may be obtained from permanent sample plots, temporary sample plots or stem analysis. Permanent sample plot information is by far the most satisfactory on which to base yield tables (Gadow and Hui 1999).

The main purpose of yield tables is to provide estimates of present yield and future increment and yield. The tables may be presented in tabular or graphical form or in the form of a regression equation relating yield to age, site and stand density. There are three main types of yield table, viz. normal, empirical and variable density (Husch 1983). Yield functions are a frequently used database for regional timber resource forecasting. Yield function estimates the production potential for a discrete number of site quality classes. Several methods for forecasting yield in the cyclic forests and plantations are available in literature.

A normal yield table is based on two independent variables, age and site (species constant), and applies to fully stocked (or normal) stands. It depicts relationships between volume/unit area together with other stand parameters and the independent variables. 'Normal' is an unfortunate term as fully stocked stands are rather unusual.

Since only two independent variables are involved, normal yield tables are conveniently constructed by graphical means. The density variable is held constant by attempting to select sample plots of a certain fixed density assessed as full (or normal) stocking. The data presented in normal yield tables are averages derived from many stands considered to be fully stocked at the time they were sampled.

In contrast to normal yield tables, empirical yield tables are based on average rather than fully stocked stands. This simplifies the selection of stands for sampling. The resulting yield tables describe stand characteristics for the average stand density encountered during the collection of field data (e.g., yield curves for *D. sissoo* in IGNP area; Fig. 4.2).

In an even-aged plantation, the available timber volume at a given age may be estimated using an empirical yield function, which is based on observations of growing stock volumes at various development stages (Gadow and Hui 1999, p. 16 & 50). Smaltschinski (1997, p 132) has provided an example of empirical yield function to model the volume in the stand as a function of site class and age.

Normal and empirical yield tables essentially have the same limitations, namely: the difficulty of locating fully stocked stands or representative average stocked stands from which to collect the basic data; stocking may not have always been 'fully stocked' or 'average'; the problem of selecting correction factors to apply to stands of density other than normal or average.

The limitations of normal and empirical yield tables led to the development of techniques for compiling tables with three independent variables, stand density being included as the third variable: hence the term variable density yield tables. Basal area/unit area, mean diameter or stand density indices are used to define the density classes (Davis 1966, Prakash 1986). Such yield tables are particularly useful for abnormal stands e.g., abnormal due to early establishment problems, insect and fungal attack, drought, fire, fluctuating demands for produce, etc. It makes up shortcoming of normal yield table and empirical yield table. Single stand growth and yield can be estimated from variable density yield table and hence it is more emphasised by the foresters (Longchang et al 1991).

Mathematical models for estimating timber yields are usually developed by fitting a suitable equation to observed data. The model is then used to predict yields for conditions resembling those of the original data set. It may be accurate for the specific conditions, but of unproven accuracy or even entirely useless in other circumstances (Gadow and Hui 1999). Thus, these models tend to be specific rather than general and require validation before applying to other areas.

Using the data from the 30 sample plots measured annually for five years, Tewari (2012) used the following equation (modified from Nagel and Kehr 1997) to develop variable density yield model for *D. sissoo* plantations in IGNP area of Rajasthan (India):

$$V = e^{a + b * \ln\left(\frac{TH}{A}\right) + c * \ln(N) + d * \ln(BA)} \qquad ...(4.16)$$

where, V = over-bark volume/ha (m^3), TH = dominant height of the trees in the stands (m), A = age of the stand (years), N = stems/ha, BA = basal area/ha (m^2).

The parameter estimates obtained were a = 2.05930, b = 0.12150, c = -0.24769, d = 1.48353. The average bias for the model was -0.084 which was only 0.08% of the mean value of the volume/ha. The average relative discrepancy between observed and predicted values was only 1.08%.

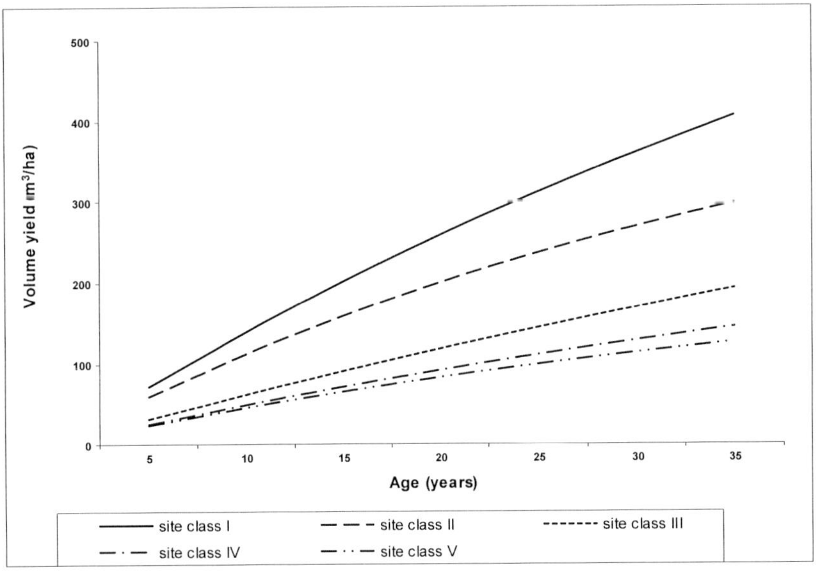

Fig. 4.2. *Volume yield vs. stand age curves for D. sissoo plantations for different site classes in IGNP area of Rajasthan*

5 Dominant Height and Site Index Model

1. HEIGHT GROWTH MODELS

Depending on the type of data available, it is possible to develop an anamorphic or a polymorphic growth model. The anamorphic approach assumes that the model parameters remain constant for any growing site, i.e., that the shape of the growth curve is independent of environmental factors. This assumption is abandoned in a polymorphic model of height growth, where the form and orientation of the growth curves with respect to the time axis is determined by the site index (Stage 1963). For any two curves, in an anamorphic family of height curves, the height of one at any age is a constant proportion of the height of the other at the same age. This proportionality does not hold in a polymorphic family of height curves. A polymorphic height model can be developed only when information is available from stem analysis or plots laid out at locations representing different site qualities.

1.1 Anamorphic Height Growth Model

To develop the anamorphic height model, the difference form equations of the Schumacher and Chapman-Richards functions may be used. The difference form of the Schumacher model (cf. Gadow and Hui 1999, p.29) assumes the following form:

$$\ln(H_2) = \ln(H_1) + b * \left(\frac{1}{t_2} - \frac{1}{t_1}\right) \qquad ...(5.1)$$

where, H_1 and H_2 are the heights corresponding to the ages t_1 and t_2.

This equation can be transformed into following equation:

$$H_2 = H_1 * \exp\left[b * \left(\frac{1}{t_2} - \frac{1}{t_1}\right)\right] \qquad ...(5.2)$$

A simple algebraic difference form of the Chapman-Richards equation (Clutter et al 1983) may be written as follows:

$$H_2 = H_1 * \left[\frac{1 - e^{-b*t_2}}{1 - e^{-b*t_1}}\right]^c \qquad ...(5.3)$$

Tewari and Gadow (2003) developed anamorphic height growth models for a pure even-aged *Prosopis cineraria* stands located in the IGNP area of Jaiselmer district in Rajasthan (India) by fitting equation (5.2) and (5.3) to the interval data. The parameter estimates and statistics are given in Table 5.1.

Table 5.1: Parameter values and statistics obtained by fitting equations (5.2) and (5.3) to the interval data of *Prosopis cineraria* stands in IGNP area of Rajasthan

Model	b	c	R^2	MSE	av. Bias
Equation 5.2	-7.5673 (0.1466)		0.931	0.12569	0.00788
Equation 5.3	0.1977 (0.0029)	2.3260 (0.1174)	0.931	0.12571	0.00819

MSE = mean squared error; in brackets standard errors for the parameter estimates are given

Tewari and Kumar (2002) also fitted the algebraic difference form of the Chapman-Richards equation (5.3) to the interval data (height-age data pairs) of stands of *Azadirachta indica* stands in Gujarat for generating anamorphic system of height growth curve. The parameter estimates and fit statistics obtained were as follows:

$$b = -0.13507, \quad c = 0.15111; \quad MSE = 0.05686, \quad R^2 = 0.994$$

2. DOMINANT HEIGHT AND SITE INDEX MODELS

Over the last few decades, biological growth functions (cf. Lundgren and Dolid 1970) have been used to describe height growth patterns. Stand growth and yield depend on the productivity of the site. A variety of methods for estimating site productivity has been developed (Clutter et al 1983, Prodan et al 1997). The most common approach involves the use of site index equations. Numerous studies have focused on the methodical aspects of constructing site index equations (e.g., Grut 1977, Ortega and Montero 1988, Zeide 1993). Site index, defined as the dominant height at a given reference age, is the most commonly used variable for evaluating forest productivity or for relating productivity to ecological variables (Clutter et al 1983, Carmean and Lenthall 1989, Payandeh and Wang 1994). Also, site index is widely used in stand growth models to estimate forest-level yields (e.g., Pretzsch 1995, Sterba 1995, Gadow and Hui 1999). Thus, the development of site index equations is a fundamental task in forest management. Many possibilities exist for generating site index equations, depending on the construction method (guide curve, parameter prediction or difference equations), the model properties, the anamorphic or polymorphic nature of the system, the number of model parameters or the presence or absence of an asymptotic height (Ortega and Montero 1988). To develop a site index equation, one needs a suitable height model. Site-index models relate height, age and site quality (potential productivity) in even-aged single-species stands and are used for predicting stand height development, and for assessing site quality.

The dominant height of a given species is more closely related to the capacity of a given site to produce wood of that species than any other one measure (Spurr and Barnes 1980). Height growth is generally more stable, directly measured stand growth statistic. Dominant height growth is independent of stocking over a fairly wide range of stand density, thus is often used as a measure of site productivity.

The most frequently listed desirable attributes of a site index system are: (1) polymorphism, (2) sigmoid (S-shaped) growth, (3) asymptote as a function of site index (increasing with increasing site index) and (4) base-age invariance (Bailey and Clutter 1974, Borders et al 1984, Scott and Voorhis 1986, Goelz and Burk 1992, Parresol and Vissage 1998). By developing a height growth model from a differential equation, these attributes may be explicitly incorporated into the equation and it may perform better outside the data used to fit the model (Pienaar and Turnbull 1973).

2.1 Polymorphic Site Index Models

Some widely used base-age variant (Ek 1971, Newnham 1988) and base-age invariant (Goelz and Burk 1992, Payandeh and Wang 1994, Clutter and Jones 1980, Korf 1939) site index models are as follows:

Ek (1971): $$H = aS^b(1-e^{-ct})^{dS^e} + \varepsilon \quad ...(5.4)$$

Newnham (1988):

$$H = as^b(1-e^{-ct})^d + \varepsilon \quad ...(5.5)$$

where, $d = \ln[S/(aS^b)]/\ln[1-\exp(-ct_1)]$; t_1 = index age

Goelz and Burk (1992):

$$H_2 = H_1\left[\frac{1-\exp\left\{-a\left(\frac{H_1}{t_1}\right)^b t_1^c t_2\right\}}{1-\exp\left\{-a\left(\frac{H_1}{t_1}\right)^b t_1^c t_1\right\}}\right]^d + \varepsilon \quad ...(5.6)$$

Payandeh and Wang (1994):

$$H_2 = aH_1^b(1-e^{-ct_2})^d + \varepsilon \quad ...(5.7)$$

where, $d=\ln[H_1/(a\,H_1^{\,b})]/\ln[1-\exp(-ct_1)]$

Clutter and Jones (1980):

$$H_2 = \exp(\exp[a\{1/t_1 - 1/t_2\}](\ln H_1 - b*1/t_1 + c) + b*1/t_2 - c) + \varepsilon \quad ...(5.8)$$

Korf (1939):

$$H_2 = b_0(H_1/b_0)^{(t_1/t_2)b_2} + \varepsilon \quad ...(5.9)$$

where, a, b, c, d, e, b_0 and b_2 are the parameters to be estimated; t_1 and t_2 are age (years) at periods 1 and 2; H_1 and H_2 are the dominant heights (m) at age t_1 & t_2; ln is the natural logarithm and ε is the error.

The accuracy and precision of the predictions depend on both, the equation and the data structure but, in general, the method involving all possible growth intervals provided most stable and consistent results (Goelz and Burk 1992, Huang 1999).

Ek and Newnham equations are base-age variant models. Ek Model is an extension of Chapman-Richards function, which account for the polymorphic growth pattern of the tree species. Newnham model is a constrained version of Ek model, which has the property of H = S at the index age i.e. for $t = t_1$. Goelz and Burk and Payandeh and Wang Models are well-behaved base-age invariant functions, which can be used to estimate height at any age given the height at any other age. Compared to Goelz and Burk model, the Payandeh and Wang model has some uniqueness in the sense that it is simpler than Goelz and Burk model because only 3 parameters need to be estimated, and unlike Goelz and Burk model, which need to be developed through the difference equation method, Payandeh and Wang model was constructed in a much simpler manner using a constraining technique. Although, the models proposed by Goelz and Burk and Payandeh and Wang are base-age invariant, their prediction accuracy may vary considerably depending on the predicted age t_1 (Goelz and Burk 1992).

Autocorrelation can be modeled as a first-order autoregressive process where the error term may be expanded to represent the autocorrelation structure inherent in fitting site index models to an all-possible growth intervals data structure (Parresol and Vissage 1998):

$$H_{ij} = f(H_i, t_i, \beta) + e_{ij} \qquad ...(5.10)$$

with $e_{ij} = \rho e_{i-1,j} + \gamma e_{i,j-1} + \varepsilon_{ij}$

where, H_{ij} represents prediction of height *i* by using height *j*, age *i* and age $j \neq i$ as a predictor variable; ρ is a parameter which accounts for the autocorrelation between the current residual and residual from estimating H_{i-1} using H_j as a predictor; γ a parameter which accounts for the autocorrelation between the current residuals and the residual from estimating H_i using H_{j-1} as a predictor; and ε_{ij} are independent and identically distributed errors.

The annual top height growth (GI) in polymorphic system can be approximated by taking the first partial derivative of the equation (5.4) with respect to age.

$$GI = \frac{\partial H}{\partial A} = b_1 S^{b_2} b_4 S^{b_5} b_3 [1 - e^{-b_3 A}]^{(b_4 S^{b_5} - 1)} e^{-b_3 A} \qquad ...(5.11)$$

Here, the parameters *a*, *b*, *c*, *d* and *e* of equation (5.4) are changed to b_1, b_2, b_3, b_4 and b_5.

The age at which this growth culminates (A_{GImax}) may be obtained by taking the first partial derivative of the equation (5.11) with respect to age and equating it to zero.

$$\frac{\partial GI}{\partial A} = b_3 e^{-b_3 A} (1 - e^{-b_3 A})^{(b_4 S^{b_5} - 1)} \left[\frac{e^{-b_3 A} (b_4 S^{b_5} - 1)}{1 - e^{-b_3 A}} - 1 \right] \qquad ...(5.12)$$

The age at which growth culminates thus obtained may be given as:

$$A_{GImax} = \frac{\ln(b_4 S^{b_5})}{b_3} \qquad ...(5.13)$$

It is to be noted that the equations given here do contain the basic limitations of the site index curves they represent.

Equation (5.4) was fitted by Tewari and Kumar (2002) on the height-age data pairs of *Azadirachta indica* stands in Gujarat having additional information of site index (index age = 20 year) to generate site index curves (Fig. 5.1) which resulted in the following parameter estimates and fit statistics:

$b_1 = 0.28720$; $b_2 = 1.77602$

$b_3 = 0.01717$; $b_4 = 0.00250$

$b_5 = 2.15350$ MSE = 0.10141, $R^2 = 0.991$

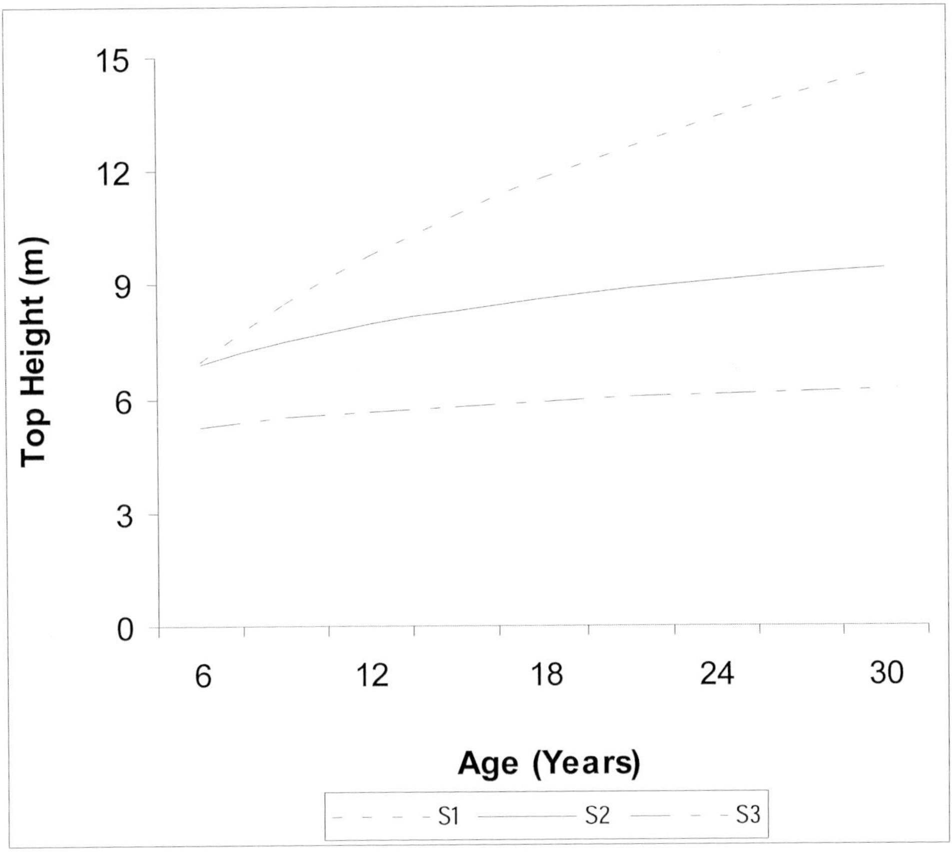

Fig. 5.1. *Top (dominant) height vs. age curves for A. indica plantations in Gujarat produced by Ek's model for different site qualities*

This provided polymorphic top height growth model. Choice of base age is typically specified so that it is less than rotation age. Ideally, base age should be chosen such that variance of the volume estimated for the forest of interest is minimized (Goelz and Burk 1992).

The above models were used to develop height growth and site index equation for *E. camaldulensis* (Tewari et al 2002), *D. sissoo* (Tewari and Kumar 2005, Fig. 5.2) and

T. undulata (Tewari and Singh 2009, Fig. 5.3) plantations in IGNP area of Rajasthan to assess site productivity, to predict dominant height at a given age and to select sites suitable for the particular species. After statistical evaluation and comparison, best models were selected which are given here.

E. camaldulensis

$$H_2 = 3.9569 H_1^{0.7584} (1 - e^{-0.0278 t_2})^d$$

where, $d = \ln[H_1/(3.9569\ H_1^{0.7584})]/\ln[1-\exp(-0.0278 t_1)]$

D. sissoo

$$H_2 = H_1 \left[\frac{1 - \exp\left\{ -0.00004 \left(\frac{H_1}{t_1} \right)^{-7.4008} t_1^{1.6645} t_2 \right\}}{1 - \exp\left\{ -0.00004 \left(\frac{H_1}{t_1} \right)^{-7.4008} t_1^{1.6645} t_1 \right\}} \right]^{0.2478}$$

T. undulata

$$H_2 = 14.9078 H_1^{-0.0539} (1 - e^{-0.0182 t_2})^d + \varepsilon$$

$$\text{where, } d = \frac{\ln\left[\frac{H_1}{14.9078 H^{-0.0539}} \right]}{\ln\left[1 - \exp\left(-0.0182 * t_1 \right) \right]}$$

An inherent drawback in using standard modeling techniques is that the full information available in the PSP data is not exploited. Usually, a model with a few independent variables is fitted to the pooled observations. A regression model fitted under these circumstances predicts the conditional mean height of the sample given the specific values of the explanatory variables. However, if there is significant variation in individual tree growth this approach will yield results different than if coefficients in a growth model are allowed to vary by individual. Consequently, it should not be assumed that the mean regression line derived from a standard approach represents the height growth development of the average trees (Biging 1985).

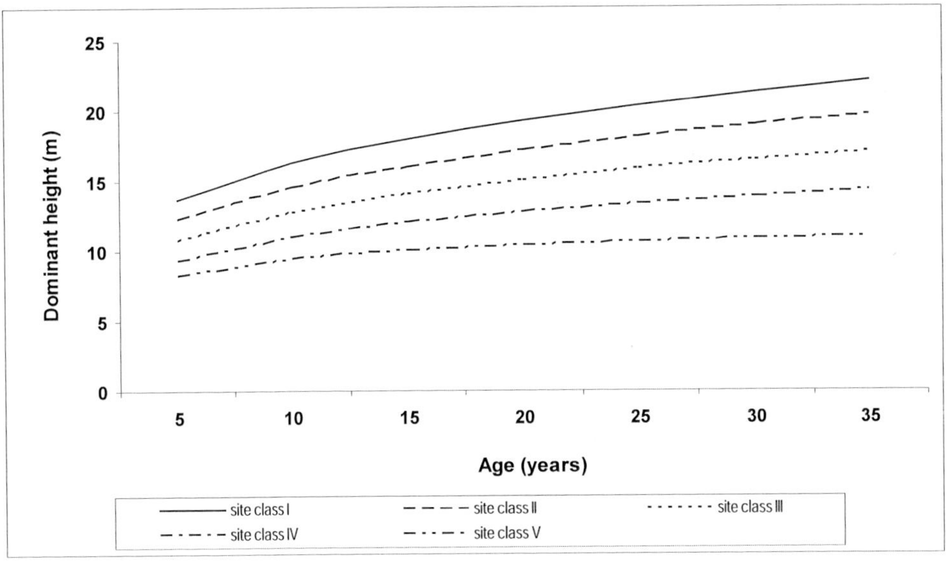

Fig. 5.2. *Height growth curves for D. sissoo plantations in IGNP area of Rajasthan for different site classes*

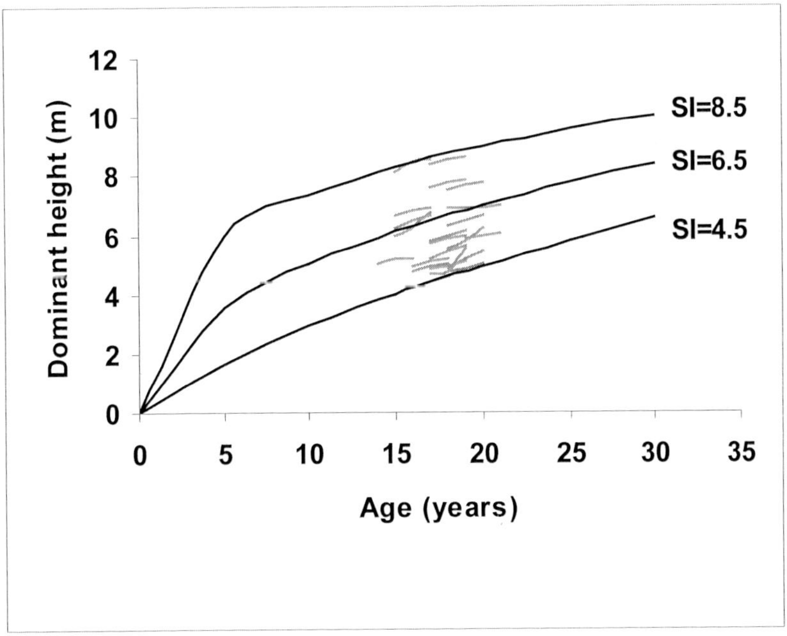

Fig. 5.3. *Site index curves for Tecomella undulata in IGNP area of Rajasthan.*

2.2 Other Algebraic Difference form of Site Index Models

Modelling height in the context of site index involves two processes: (1) estimating site index at base age given height at some other age, and (2) estimating height at some desired age given site index (Goelz and Burk 1992, 1996). These two processes may be modeled by individual functions for each process. However, in this case, height is assumed to be measured with error when it appears on the left-hand side of the equation, but is fixed when it appears on the right-hand side. This assumption causes a bias in the site index system and neither the height prediction equation nor the site index prediction represents the true functional relationship between height and age across levels of site index (Parresol and Vissage 1998). To avoid parameter bias and simultaneously optimize the two processes, the fitting of an algebraic difference equation (ADE) is recommended (Goelz and Burk 1992, 1996). Some of the recent widely used polymorphic algebraic difference equations, derived from the base model Champan-Richards (Chapman 1961, Richards 1959), Korf (1939), Hossfeld (1882) and Sloboda (1971), are given in Table 5.2.

Table 5.2. Some important algebraic difference equations for modelling site index

Algebraic difference equation	Base model
$H_2 = b_0(H_1/b_0)^{\ln[1-\exp(-b_1t_2)]/\ln[1-\exp(-b_1t_1)]}$	Chapman-Richards $H = b_0[1-\exp(-b_1t)]^{b_2}$ solved by b_2
$H_2 = b_0\left(1-\left[1-(H_1/b_0)^{1/b_2}\right]^{t_2/t_1}\right)^{b_2}$	Chapman-Richards $H = b_0[1-\exp(-b_1t)]^{b_2}$ solved by b_1
$H_2 = b_0(H_1/b_0)^{(t_1/t_2)b_2}$	Korf $H = b_0\exp(-b_1/t^{b_2})$ solved by b_1
$H_2 = b_0\exp(-b_1/t_2^z)$ with $z = \ln[-b_1/\ln(H_1/b_0)]/\ln(t_1)$	Korf $H = b_0\exp(-b_1/t^{b_2})$ solved by b_2
$H_2 = (H_1+d+r)/[2+(4\beta/t_2^{b_2})/(H_1-d+r)]$ with $d = \beta/Asi^{b_2}$ and $r = \sqrt{(H_1-d)^2+4\beta H_1/t_1^{b_2}}$	Hossfeld IV $H = b_0/(1+b_1/t^{b_2})$ solved by b_0 and assuming $b_1 = \beta/S$
$H_2 = b_0(H_1/b_0)^{\exp(Z)}$ with $Z = b_2/(b_3-1)t_2^{(b_3-1)} - b_2/(b_3-1)t^{(b_3-1)}$	Sloboda $H = b_0\exp(-b_1\exp[b_2/(b_3-1)t^{(b_3-1)}])$ solved by b_1
$H_2 = H_1([1-\exp(-zt_2)]/[1-\exp(-zt_1)])^{b_2}$ with $z = b_3(H_1/t_1)^{b_4}t_1^{b_5}$	Chapman-Richards $H = b_0[1-\exp(-b_1t)]^{b_2}$ solved by b_0 and assuming $b_1 = b_3(H_1/t_1)^{b_4}t^{b_5}$

Here, H_1 is the dominant height (m) at age t_1 (years); H_2 the dominant height (m) at age t_2 (years); Asi an age ranged from 5 to 20 years to reduce the mean squared error; ln the natural logarithm; S the site index at the reference age and β, b_0, b_1, b_2, b_3, b_4 and b_5 are the parameters to be estimated.

Height growth and site index equation were developed for *A. nilotica* and *E.* hybrid (Tewari et al 2007, Fig. 5.4) plantations in Gujarat using the equations given in Table 5.1 which are as below:

E. hybrid

$$H_2 = 130.2464(H_1 / 130.2464)^{\exp(Z)}$$

with, $Z = 0.1098 / (0.5129 - 1)A_2^{(b_3-1)} - 0.1098 / (0.5129 - 1)A_1^{(0.5129-1)}$

A. nilotica

$$H_2 = 58.4434(H_1 / 58.4434)^{\exp(Z)}$$

with, $Z = 0.0635 / (0.3073 - 1)A_2^{(b_3-1)} - 0.0635 / (0.3073 - 1)A_1^{(0.3073-1)}$

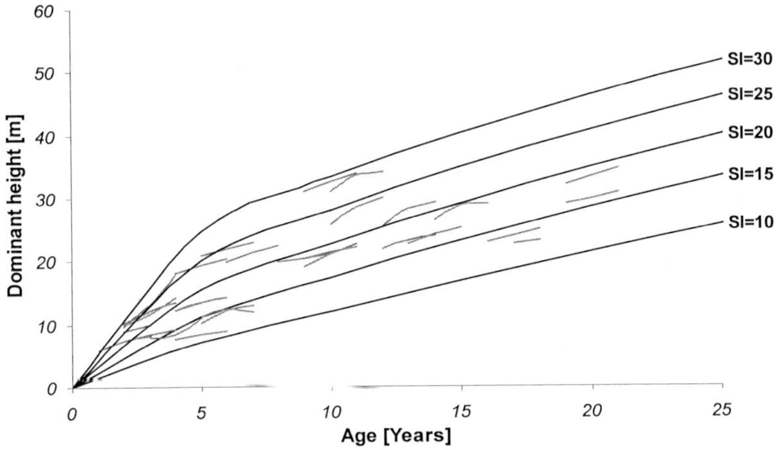

(a)

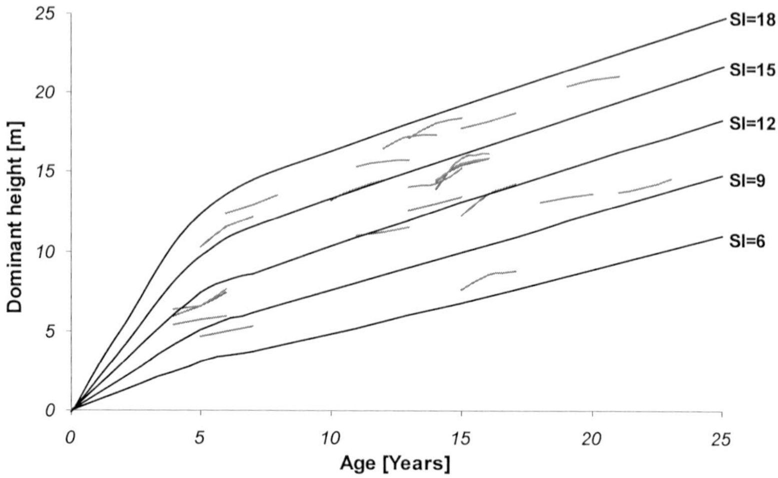

(b)

Fig. 5.4. *Site index curves generated for (a) E. hybrid and (b) A. nilotica in Gujarat*

2.3 Dynamic Site Index Models Using GADA

Bailey and Clutter (1974) used the 'Algebraic Difference Approach (ADA)' for incorporating the base-age invariance feature for dynamic site index models. They emphasized that the predictions of a site index model should not be affected by the selection of a base-age within a single equation. This base-age invariance feature has been a specialized dynamic property in site index models in order to offer some solutions for predictions with different base-ages (Cieszewski 2003). The main shortcoming of the ADA approach is the development of a polymorphic site index model with single asymptotes for all sites, and hence there is a need for a polymorphic site index model which has multiple asymptotes. Thus, Cieszewski and Bailey (2000) further expanded the algebraic difference parameter prediction approach (ADA) and introduced the 'Generalized Algebraic Difference Approach (GADA)'. The GADA method can be used to develop base-age invariant models with variable asymptotes, which are capable of describing a polymorphic family of growth curves (Wang et al 2008). The principal advantage of GADA models is the fact that important theories about the site index models are implemented, including asymptotic growth, polymorphism and base-age invariance. It is has been reported that the dynamic site index models derived by GADA have shown better and more consistent properties than the fixed base-age site index models (Cieszewski 2001, 2002, 2003, Cieszewski and Nigh 2002, Trincado et al 2003, Diéguez-Aranda et al 2005, Nord-Larsen 2006, Adame et al 2006, Bravo-Oviedo et al 2007, Cieszewski and Strub 2008).

A rather large number of mathematical equations can be used to describe dominant height growth (see for example the 74 equations documented by Kiviste et al 2002). Out of them, three well-known growth functions that have been found to perform particularly well in previous work: 1. the Korf (1939, also cited in Lundqvist 1957), 2. the Hossfeld (Hossfeld 1822) and 3. the Bertalanffy-Richards (Bertalanffy 1949, 1957, Richards 1959) models.

On the basis of these equations, several dynamic models were formulated using the GADA for developing the projection function. Nowadays it is believed that anamorphic curves do not adequately represent the growth of the main tree or stand variables and therefore, only the possible polymorphic solutions of the above mentioned equations were considered for analysis. Some of these solutions were earlier discarded because the fitting curves did not describe the observed trends in the data well. The formulations of six dynamic equations, are presented in Table 5.3. All of the equations are base-age invariant.

As general notational convention, $a_1, a_2,\ldots, a_n$ were used to denote parameters in base models, while $b_1, b_2,\ldots, b_m$ were used for global parameters in subsequent GADA formulations. All the GADA based models have the general implicit form of:

$$Y = f(t, t_0, Y_0, b_1, b_2 \ldots b_m) \qquad \ldots(5.14)$$

Models E1, E3 and E5 were derived by applying the GADA method to the Korf, Hossfeld and Bertalanffy-Richards functions, respectively, by considering only parameter a_2 to be site specific. In this case the GADA method is equivalent to the ADA method. All these models are polymorphic with a single asymptote.

Dynamic equations E2, E4 and E6 were developed by considering two parameters to be site-specific. Model E2 was derived by Cieszewski (2004) on the basis of the Korf function by considering both parameters a_1 and a_2 to be dependent on X. To facilitate such derivation, the base equation was re-parameterized into a more suitable form for manipulation of these two parameters, using e^X instead of a_1. Parameter a_2 was expressed as a linear function of the inverse of X. Model E4 was derived by Cieszewski (2002) from the Hossfeld function, by replacing a_1 with a constant plus the unobserved site variable X, and a_2 by b_2/X. Model E6 was developed by Krumland and Eng (2005) by expressing the asymptote as an exponential function of X and the shape parameter as a linear function of the inverse of X.

The parameter estimation of the regression equations describing the behaviour of individuals over time often has associated problems of serial correlation (i.e., correlation between the residuals within the same individual). To account for this possible autocorrelation, the error term may be modelled using a first-order continuous autoregressive error structure (CAR (1)) that allows the models to be applied to irregularly spaced, unbalanced data (Gregoire et al 1995, Zimmerman and Núñez-Antón 2001). The CAR (1) expands the error terms in the following way:

$$e_{ij} = \psi_1 \rho^{t_{ij}-t_{ij-1}} e_{ij-1} + \varepsilon_{ij} \qquad \ldots(5.15)$$

where e_{ij} is the *j*th ordinary residual on the *i*th individual (i.e., the difference between the observed and the estimated dominant height of plot *i* at age measurement *j*), $\psi_1=1$ for $j>1$ and it is zero for $j=1$, ρ is the first-order autoregressive parameter to be estimated, $t_{ij}-t_{ij-1}$ is the time distance separating the *j*th from the *j*th-1 observations within each individual ($t_{ij}>t_{ij-1}$) and ε_{ij} is now the error term under conditions of independence and no heteroscedasticity.

Height growth and site index equation were developed for teak in Karnataka (Tewari et al 2014a, Fig. 5.5) using the GADA equations given in Table 5.3. After considering the statistical criteria and analysing the graphical output, model E2, derived from the Korf equation, was proposed for height growth prediction and site classification of Teak in Karnataka (India) and the expression is given below:

$$H_2 = \exp(X_0)\exp)(-(13.6964 + 22.3517/X_0)t_2^{-1.2229}) \text{ with}$$

$$X_0 = \frac{1}{2}t_1^{-1.2229}(13.6964 + t_1^{1.2229}\ln(H_1) \pm \sqrt{4 \cdot 22.3517t_1^{1.2229} + (-13.6964 - t_1^{1.2229}\ln(H_1))^2})$$

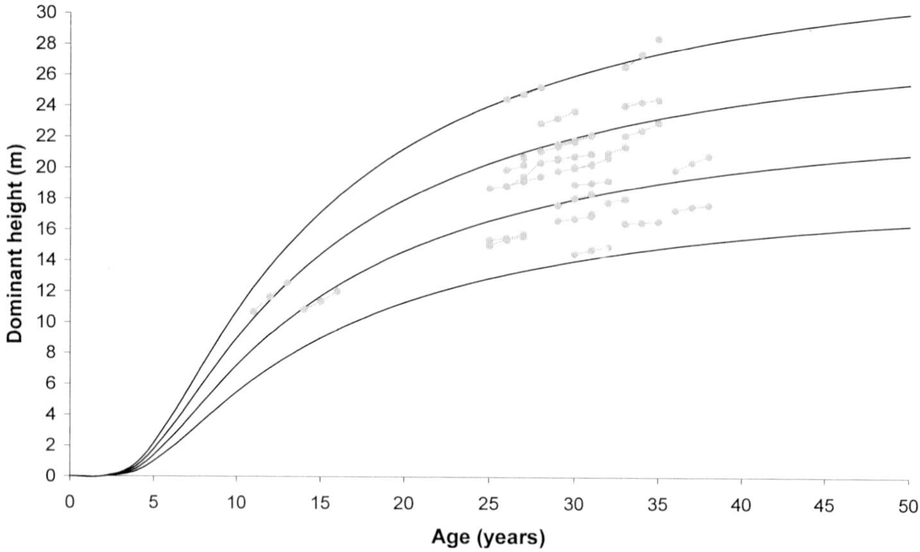

Fig. 5.5. *Dominant height curves for teak plantations in Karnataka for site indices of 14, 18, 22 and 26 m at a reference age of 30 years based on GADA based Korf model.*

Table 5.3. Some dynamic site index equations derived using Generalized Algebraic Difference Approach (GADA)

Base equation	Parameter related to site	Solution for X with initial values (t_0, Y_0)	Dynamic equation	Equation no.
Korf: $Y = a_1 \exp(-a_2 t^{-a_3})$	$a_1 = X$	$X_0 = Y_0 / \exp(-b_1 \cdot t_0^{-b_2})$	$Y = Y_0 \cdot \exp(-b_1 \cdot (t_1^{-b_2} - t_0^{-b_2}))$	E1
	$a_1 = \exp(X)$ $a_2 = b_1 + b_2/X$	$X_0 = \frac{1}{2} t_0^{-b_3} \left(b_1 + t_0^{b_3} \ln(Y_0) \pm \sqrt{4 b_2 t_0^{b_3} + \left(-b_1 - t_0^{b_3} \ln(Y_0)\right)^2} \right)$	$Y = \exp(X_0) \exp(-(b_1 + b_2/X_0) t^{-b_3})$	E2
Hossfeld: $Y = \dfrac{a_1}{1 + a_2 t^{-a_3}}$	$a_1 = X$	$X_0 = Y_0 \cdot (1 + b_1 \cdot t_0^{-b_2})$	$Y = Y_0 \cdot \dfrac{1 + b_1 \cdot t_0^{-b_2}}{1 + b_1 \cdot t_1^{-b_2}}$	E3
	$a_1 = b_1 + X$ $a_2 = b_2/X$	$X_0 = \frac{1}{2}\left(Y_0 - b_1 \pm \sqrt{(Y_0 - b_1)^2 + 4 b_2 Y_0 t_0^{-b_3}} \right)$	$Y = \dfrac{b_1 + X_0}{1 + b_2 / X_0 t^{-b_3}}$	E4
	$a_1 = X$	$X_0 = Y_0 / (1 - \exp(-b_1 \cdot t_0))^{b_2}$	$Y = Y_0 \left(\dfrac{1 - \exp(-b_1 t)}{1 - \exp(-b_1 t_0)} \right)^{b_2}$	E5
Bertalanffy-Richards: $Y = a_1 (1 - \exp(-a_2 t))^{a_3}$	$a_2 = X$	$X_0 = -\ln(1 - (Y_0 / b_1)^{1/b_3}) / t_2$	$Y = b_1 \left(1 - \left(1 - \left(\dfrac{Y_0}{b_1} \right)^{1/b_3} \right)^{t/t_0} \right)^{b_3}$	E6
	$a_1 = \exp(X)$ $a_3 = b_2 + b_3/X$	$X_0 = \frac{1}{2}\left((\ln Y_0 - b_2 L_0) \pm \sqrt{(\ln Y_0 - b_2 L_0)^2 - 4 b_3 L_0} \right)$ with $L_0 = \ln(1 - \exp(-b_1 t_0))$	$Y = Y_0 \left(\dfrac{1 - \exp(-b_1 t)}{1 - \exp(-b_1 t_0)} \right)^{(b_2 + b_3/X_0)}$	E7

6

Stand Density and Basal Area Development

1. STAND DENSITY

The term stand density and stocking are frequently erroneously used interchangeably, yet they are not synonyms. Stand density is a measure of the number of trees on a given piece of land whereas stocking refers to the degree of site occupancy. If there were no mortality within a stand of trees, stand density would be constant while stocking would continually change as the trees grew in size (Gadow and Bredenkamp 1992). Stands not at optimal stocking do not make the best use of the site and the regulation of stand density is therefore an important function of timber management.

Estimating the potential density of forest stands, in terms of the surviving trees per ha, is a central element of growth modelling. It is also one of the most difficult problems to solve, mainly because suitable data from untreated, densely stocked stands are rarely found.

1.1 Limiting Line

Populations of trees growing at high densities are subject to density-dependent mortality or self-thinning (Westoby 1984). An important empirical size-density relationship which has received much attention in the literature is Yoda's 3/2 power law (Yoda et al 1963). For a given average tree size, there is a limit to the number of trees per ha that may co-exist in an even-aged stand. The relationship between the average tree size and the number of live trees per unit area may be described by means of a limiting line. A convenient model for this limiting relationship is (Reineke 1933):

$$N_{max} = \alpha_0 Dg^{\alpha_1} \quad ...(6.1)$$

with N_{max} = maximum number of surviving trees per ha,

Dg = quadratic mean diameter [cm]

α_0, α_1 = empirical parameters

1.2 Estimation of Potential Density

The relationship between the quadratic mean diameter (Dg), the dominant stand height (Ho) and the number of stems per unit area (N) can be described using a hyperbola (Goulding 1972):

$$Dg = \frac{1}{\alpha_0 H^{\alpha_1} N + \beta_0 H^{\beta_1}} \quad ...(6.2)$$

The per-hectare basal area is equal to

$$G = \frac{\pi}{4} Dg^2 \cdot N = \frac{\pi \cdot N}{4[\alpha_0 H^{\alpha_1} N + \beta_0 H^{\beta_1}]^2} \quad ...(6.3)$$

Upon setting the first derivative of the equation (6.3) with respect to N equal to zero, the stems per ha at maximum basal area are obtained (Sterba 1975):

$$N_{G\max} = \frac{\beta_0}{\alpha_0} H^{(\beta_1 - \alpha_1)} \quad ...(6.4)$$

Substituting N in equation (6.2), this equation gives the quadratic mean diameter at maximum basal area:

$$Dg_{G\max} = \frac{1}{2\beta_0 H^{\beta_1}} \quad ...(6.5)$$

Solving the equation (6.5) for H (Sterba 1987) and substituting this expression in equation (6.4), finally one can obtain:

$$N_{G\max} = \frac{\beta_0}{\alpha_0} \cdot (2\beta_0)^{\frac{\alpha_1}{\beta_1} - 1} Dg_{G\max}^{\frac{\alpha_1}{\beta_1} - 1} \quad ...(6.6)$$

The resulting equation represents the limiting line.

This technique for determining the limiting line is suitable for estimating potential density with data from unthinned stands. However, the data set should include stands where self-thinning has occurred otherwise the slope of limiting line may be estimated incorrectly (Gadow and Hui 1999).

Using the above method, limiting line has been determined to estimate potential density in *E. camaldulensis* (Tewari 2007c), *D. sissoo* (Tewari and Gadow 2008) and *T. undulata* (Tewari 2010) plantations in IGNP area of Rajasthan and *A. indica* (Tewari 2004), *E.* hybrid (Tewari and Singh 2008) and *A. nilotica* (Tewari 2008) plantations in Gujarat. The equations develop to generate limiting line for these species are given below:

E. camaldulensis

$$N_{G\max} = 35835.69 D_{G\max}^{-1.1905}$$

D. sissoo

$$N_{G\max} = 68761.87 D_{G\max}^{-1.5109}$$

T. undulata

$$N_{G\max} = 57629.65 D_{G\max}^{-1.46966}$$

A. indica

$$N_{G\max} = 54905.39 D_{G\max}^{-1.48687}$$

E. hybrid

$$N_{G\max} = 29443.7 D_{G\max}^{-1.16145}$$

A. nilotica

$$N_{G\max} = 23972.26 D_{G\max}^{-1.31384}$$

The relationship between the quadratic mean diameter and the number of living trees per unit area along with the limiting line is shown in Fig. 6.1. The solid line in the figure represents the potential density in the plots, i.e., the maximum stems ha^{-1} the plots can have with respect to the quadratic mean diameter at maximum basal area.

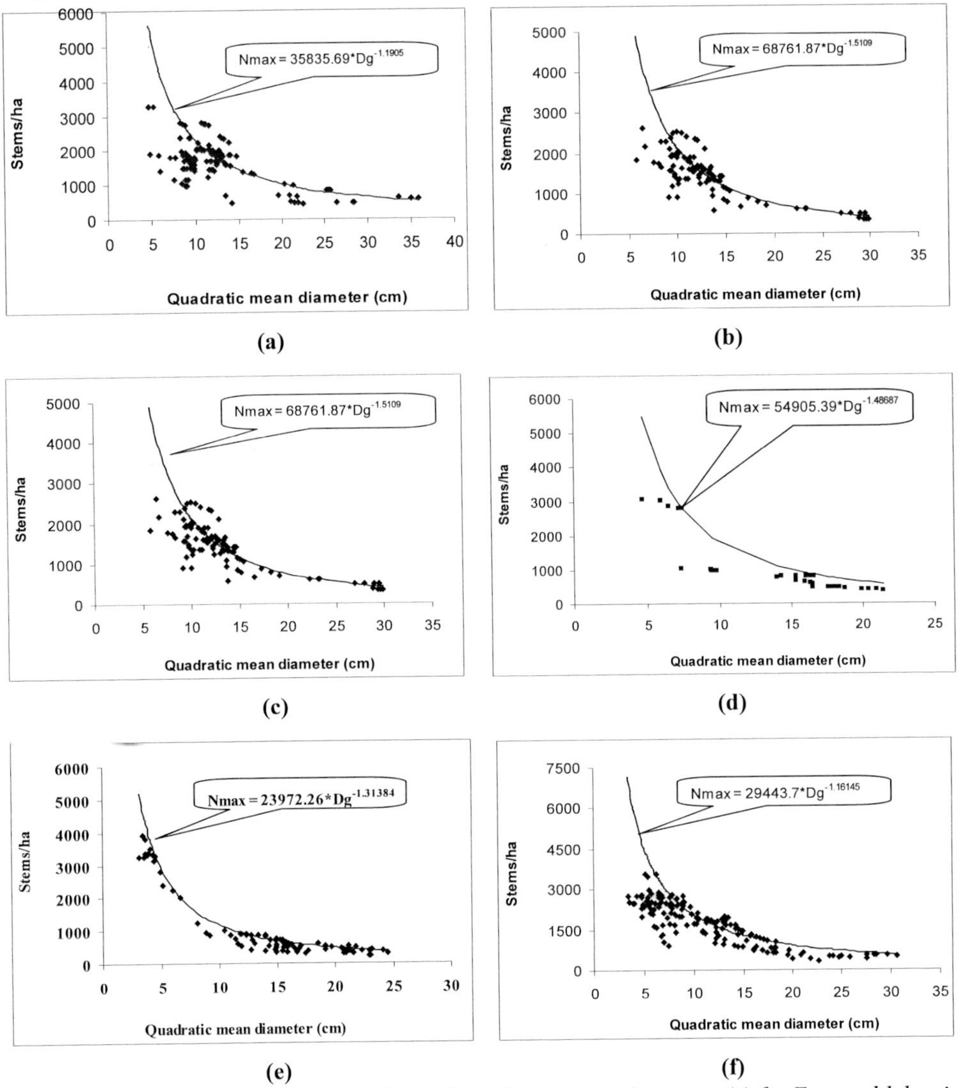

Fig. 6.1. *Relationship between stems/ha and quadratic mean diameter; (a) for E. camaldulensis, (b) for D. sissoo, (c) for T. undulata, (d) for neem, (e) for A. nilotica and (f) for E. hybrid. The solid line is limiting line. The equation is derived from equation 6.6.*

1.3 Mortality

The natural decline of the number of surviving trees in an unthinned forest is usually characterised by intermittent brief spells of high mortality, followed by long periods of low mortality (Gadow and Hui 1999), which makes estimation of tree survival a real challenge.

In the practice of forest management, the limiting line is hardly ever reached (Gadow and Bredenkamp 1992, p. 75). However, this does not mean that trees do not die before the limit is reached. The results from spacing trials show that mortality processes are effective well before the maximum basal area is attained.

Clutter and Jones (1980) used a differential equation for modelling the rate of change in the number of stems per ha (N) as a function of stand age:

$$\frac{1}{N}\frac{\partial N}{\partial t} = \alpha \cdot N^{\beta} \cdot t^{\gamma} \qquad ...(6.7)$$

with N = number of surviving trees per ha

t = stand age [years] and α, β, γ = empirical parameters

It is convenient, after integrating, to use the following algebraic difference form:

$$N_2 = [N_1^a + b(t_1^c - t_2^c)]^{\frac{1}{a}} \qquad ...(6.8)$$

with N_1, N_2 = stems per ha corresponding to stand ages t_1 and t_2

a, b, c = empirical parameters

A slightly extended version of this equation, including site index as an independent variable, will be as follows:

$$N_2 = \left[N_1^{\,a} + \left(b + \frac{c}{SI} \right) \left(\left[\frac{t_2}{10} \right]^d - \left[\frac{t_1}{10} \right]^d \right) \right]^{\frac{1}{a}} \qquad ...(6.9)$$

with N_1, N_2 = stems per ha corresponding to stand ages t_1 and t_2

SI = site index

a, b, c, d = empirical parameters

The natural decline of stem numbers has been modelled for the pure even-aged stands of *E. camaldulensins* (Tewari 2007c), *D. sissoo* (Tewari and Gadow 2008) plantations in IGNP area of Rajasthan and *A. nilotica* (Tewari 2004) plantations in Gujarat. The equations developed for estimating the rate of change in the number of stems per ha in the stands are:

E. camaldulensis

$$N_2 = \left[N_1^{\,0.5027} + \left(-3.0526 + \frac{21.3766}{SI} \right) \left(\left[\frac{A_2}{10} \right]^{2.4940} - \left[\frac{A_1}{10} \right]^{2.4940} \right) \right]^{\frac{1}{0.5027}}$$

D. sissoo

$$N_2 = [N_1^{0.14148} + 0.00686(A_1^{1.27431} - A_2^{1.27431})]^{\frac{1}{0.14148}}$$

A. indica

$$N_2 = \left[N_1^{0.000033} + \left(-0.000012 + \frac{0.000046}{SI} \right) \left(\left[\frac{A_2}{10} \right]^{0.956197} - \left[\frac{A_1}{10} \right]^{0.956197} \right) \right]^{\frac{1}{0.000033}}$$

2. STAND DENSITY MANAGEMENT DIAGRAM

Stand density management involves the manipulation and control of growing stock through initial spacing or subsequent thinning to achieve specific management objectives (Newton 1997, Newton et al 2005). Manipulation of stand density with thinning affects stand structure, tree size yield, and optimal rotation length. It is probably the single most influential activity a silviculturist can perform between successive regeneration periods, and is often the best means for achieving a wide range of stand management objectives (Long 1985). Adequate stand density levels determination is a difficult task that involves biological, technological, economical and operational factors driven by species, site and management frameworks (Valbuena et al 2008). The process requires selection of upper and lower limits of growing stock while taking into account the fact that the upper limit is chosen to ensure site occupancy, to maintain adequate stand-level yields and to produce trees of a desired form, whereas the lower limit is chosen to promote individual-tree growth rates (Dean and Baldwin 1996).

Stand density management diagrams (SDMD) graphically represent the relations between stand density, height, diameter and volume for even-aged stands (Archibald and Bowling 1995, Newton 1997, Barrio and Álvarez-González 2005). These diagrams are based on the self-thinning rule and the relationships between diameter, dominant height, density and volume for even-aged stands (Archibald and Bowling 1995). In fact, these diagrams can be used to predict size-density trajectories under different silvicultural strategies (Newton 1997). SDMDs effectively translate general management objectives into stand-specific thinning prescriptions (Castedo-Dorado et al 2009). The apparent simplicity of SDMDs belies their ability to display the complex dimensional relationships of developing stands (Kumar et al 1995). Essentially, SDMDs are simple stand average models which graphically characterize yield, density and mortality at various stages of stand development (Newton and Weetman 1994) and are useful in designing alternative density management regimes, in displaying the consequences of stand density manipulation and in translating general objectives into practical thinning prescriptions (Long et al 1988). They have also been used to develop regimes relating to the manipulation of wildlife habitat (Smith and Long 1987, Lilieholm et al 1994).

The development of SDMDs has greatly facilitated stand density management decision making. Their utility has been largely limited to evaluating density management outcomes in terms of mean tree size and stand-level volume yields

(Newton 1997). Recent innovations include the addition of structural yield prediction in the development of SDMDs (Newton et al 2005) and of traditional stocking guides (Gove 2004), which enable forest managers to estimate the number of trees in each diameter class at any point during stand development.The use of SDMDs is one of the most effective methods for the design, display and evaluation of alternative density management regimes in even-aged stands (Jack and Long 1996). The diagrams are constructed by characterization of the growing stock with indices that relate the average tree size to density. Several density indices have been used e.g. the stand density index (Reineke 1933), the self-thinning rule (Yoda et al 1963), the relative density index (Drew and Flewelling 1979) and the relative spacing index (Wilson 1946). The great advantage of these density indices is that they are independent of site quality and stand age (Long 1985, McCarter and Long 1986).

2.1 Construction of the SDMD

Structurally, SDMDs consist of a number of functional and empirical quantitative relationships that collectively represent the cumulative effect of various underlying competition processes on tree and stand yield variables (Newton et al 2005). The relative spacing index (RS, also known as Hart or Hart–Becking index) may be used to characterize the growing stock level (Clutter et al 1983, Gadow and Hui 1999). This index was calculated as the ratio of the average distance between trees to the average dominant height of the stand, expressed as a percentage. Assuming a square spacing, it can be written as follows:

$$RS(\%) = \frac{100}{\sqrt{N}H_{100}}100 \qquad ...(6.10)$$

RS is particularly useful for characterizing the growing stock levels in SDMDs for several reasons (Castedo-Dorado et al 2009): (1) it is independent of site quality and stand age except for very young stands, (2) dominant height is, from a biological point of view, the best index for establishing thinning intervals, (3) the linkage between dominant height growth and forest production adds further utility to these diagrams for forest management purposes, and (4) it is sometimes used as guide to control density in intensively managed plantations.

Tewari and Álvarez-González (2014) developed stand density management diagram for forest teak plantations in Karnataka (India). The following two equations were fitted using non-linear regression in the construction of the stand density management diagram:

$$d_g = b_0 \times N^{b_1} \times H_{100}^{b_2} \qquad ...(6.11)$$

$$V = b_3 \times d_g^{\ b_4} \times H_{100}^{b_5} \times N^{b_6} \qquad ...(6.12)$$

where, N = number of stems/ha, H_{100} = dominant height, V = total stand volume, d_g = quadratic mean diameter

Total stand volume was calculated from a compatible tree volume equation developed by Tewari et al (2013). The first equation relates quadratic mean diameter to number of trees per ha and dominant height. This equation is based on the relationship among average tree size, density and an indicator of productivity (Goulding 1972, Long and Shaw 2005). The second equation relates the current stand volume to the quadratic mean diameter, dominant height and the number of trees per ha (McCarter and Long 1986, Dean and Baldwin 1993, Luis and Fonseca 2004).

Eqs. (6.11) and (6.12) together define a structurally simultaneous system of equations, where N and H_{100} are the exogenous variables (variables that appear only on the right-hand sides of the equations and whose values are determined independently of the system); V is an endogenous variable (a variable that the model is intended to explain or predict and that only appears on the left-hand side of an equation); and d_g is an endogenous instrumental variable (endogenous variables also appear on the right-hand side of equations). Since the error components of the variables on the left-hand side and the right-hand side are correlated, the full information maximum likelihood (FIML) technique was applied to fit both equations simultaneously using the MODEL procedure of SAS/ETS (SAS Institute Inc. 2004).

The isolines for the growing stock level (expressed by the RS) and for the stand variables included in the stand-level model (quadratic mean diameter and stand volume), were obtained by solving for N in equations (6.10-6.12) through a range of H_{100} and setting RS, d_g and V constant, respectively. The values of all the variables used to develop the SDMD ranged between the maximum and minimum values observed for teak in Karnataka (Table 6.1).

Table 6.1: Parameter coefficients and fit statistics obtained by fitting simultaneously equations (6.11 and 6.12) predicting quadratic mean diameter and stand volume

Coefficients	Values	Aprox. SEE	R^2	RMSE
b_0	34.57	12.5405	0.7421	2.30
b_1	-0.3646	0.0406		
b_2	0.6165	0.0680		
b_3	0.000043	0.0000037	0.9808	11.66
b_4	2.2652	0.2615		
b_5	0.6818	0.2041		
b_6	0.9886	0.0948		

SEE = Standard Error of Estimate, RMSE=Root Mean Squared Error

A mortality model may also be considered in the representation of density management regimes in the diagram. The mortality model used by Tewari and Álvarez-González (2014) is derived from the following differential equation:

$$\frac{1}{N}\cdot\frac{\Delta N}{\Delta t} = \alpha \cdot N^{\beta} \cdot \delta^{t} \qquad ...(6.13)$$

It is assumed that the rate of change of stand density depends on a power function of the initial stand density and an exponential function of age. This differential equation has two different solutions depending on the value of β.

$$\frac{\Delta N}{N^{\beta+1}} = \alpha \cdot \delta^{t} \cdot \Delta t \Rightarrow \begin{cases} \beta \neq 0 \quad N^{-\beta} = \alpha \cdot \delta^{t} \cdot \dfrac{1}{\log \delta} + C \\ \beta = 0 \quad \log N = \alpha \cdot \delta^{t} \cdot \dfrac{1}{\log \delta} + C \end{cases} \qquad ...(6.14)$$

The first solution gives the following algebraic difference equation by substituting C:

$$\begin{cases} N_1^{-\beta} = \alpha \cdot \delta^{t_1} \cdot \dfrac{1}{\log \delta} + C \\ N_2^{-\beta} = \alpha \cdot \delta^{t_2} \cdot \dfrac{1}{\log \delta} + C \end{cases} \Rightarrow N_2^{-\beta} = \alpha \cdot \delta^{t_2} \cdot \frac{1}{\log \delta} + N_1^{-\beta} - \alpha \cdot \delta^{t_1} \cdot \frac{1}{\log \delta}$$

$$\Rightarrow \qquad N_2 = \left(N_1^{-\beta} + \alpha \cdot \frac{1}{\log \delta} \cdot [\delta^{t_2} - \delta^{t_1}] \right)^{1/-\beta}$$

$$N_2 = (N_1^{b_8} + b_9 \cdot [\delta^{t_2} - \delta^{t_1}])^{1/b_7} \text{ with.. } b_9 = \alpha \cdot \frac{1}{\log \delta} \text{ ..and.. } b_8 = -\beta$$

The following model was developed which enables prediction of the reduction in tree number due to density-dependent mortality:

$$N_2 = [N_1^{0.0014} + 0.99998^{A_2 - 100} - 0.99998^{A_1 - 100}]^{\frac{1}{0.0014}} \qquad ...(6.15)$$

where N_1 and A_1 represent the predictor number of trees per ha and age (years), respectively, and N_2 is the predicted number of trees per ha at age A_2.

This equation is sensitive only to initial stand density and stand age. Since equation (6.15) depends on stand age, the site quality system (Tewari et al 2014) was used to convert the age variable to dominant height (given a specified site index), which can be directly represented in the SDMD. A SDMD was developed by superimposing the expected size-density trajectories, i.e. the values of RS, the isolines for quadratic mean diameter and the isolines for stand volume on the bivariate graph (Fig. 6.2).

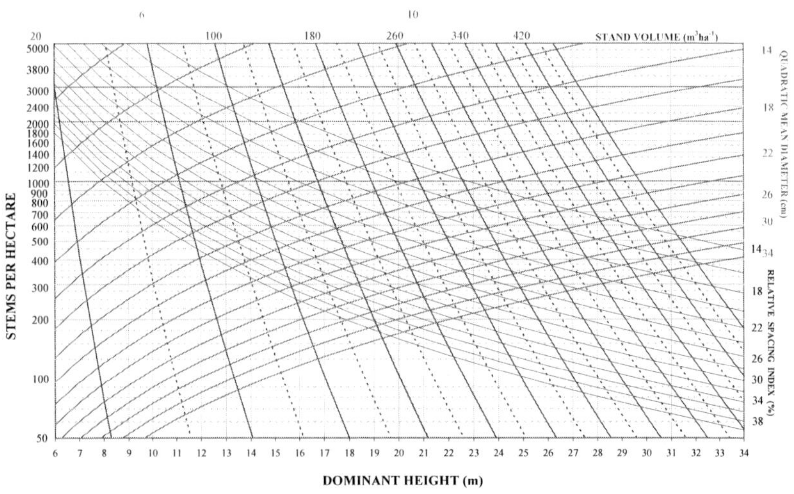

Fig. 6.2. *The Stand Density Diagram for teak stands in Karnataka (India)*

The dimensional relationships represented by the diagram (Fig. 6.2) are presumed to be largely independent of both site quality and stand age. For example, a stand with a d_g of 20 cm and 1000 trees/ha would have the same H_1 and total volume, be at the same stage of stand development, and have the same growth-growing stock relations whether it was a relatively young stand on a high-quality site or an older stand on a low-quality site. The utility of the diagram at a particular locale is enhanced if information on the dominant height-age relation makes it possible to estimate rotation length and ages of intermediate stand entries.

2.2 Potential uses and Limitations of SDMDs

The SDMD has a wide potential use because the data required for the equations and diagrams are usually available from common timber inventories. The SDMD may be useful for the rapid analysis of many alternative density management regimes; when one or few alternatives are identified, however, if more detailed yield analysis is required it should be provided from a dynamic stand growth model predicting rates of change under different management approaches and stand conditions (Castedo-Dorado et al 2009).

Although different thinning regimes can be analysed using the SDMD, the lack of information from thinned plots for developing the SDMD limits its use. Furthermore, silvicultural treatments can induce the so-called "memory problem" (Drew and Flewelling 1979, Long 1985, Jack and Long 1996) associated with changes in stand structure over time; i.e. a stand which was not thinned will not have the same average size or the same allometric relationships as a comparable stand of the same density that was heavily thinned.

In the use of SDMDs, it is assumed that such structural differences are either short-lived or of limited silvicultural importance (Drew and Flewelling 1979). Therefore, these limitations should be considered before applying the SDMD and, according to Newton (2003), further information about the effects of different thinning treatments on the residual stand structure and the temporal variation in thinning responses is required.

There are limitations to SDMDs since these are only average stand models. These diagrams do not provide any information on the range of diameter and height classes within a stand. As a result, a SDMD cannot be used as a yield table. Volume estimates derived from a SDMD are gross approximations and are only appropriate for comparative evaluations of crop planning scenarios (Archibald and Bowling 1995).

3. BASAL AREA PREDICTION MODEL

An important element of a stand growth model is the quadratic mean diameter. This quantity, which is essential for estimating product yields, may be derived from the per-hectare basal area and the number of stems per hectare. Modelling its growth usually proves to be surprisingly easy, given data from permanent plots grown at different densities. However, problems usually arise when the model is applied in thinned stands, where the mean diameter may develop in totally different ways, depending on the type of thinning.

Area-based variables entail the greatest potential for error. For this reason, it is preferable, though more difficult, to model basal area and number of stems, and derive the quadratic mean diameter from these quantities using $Dg = \sqrt{40,000 / \pi \cdot G/N}$.

In stands of the same age and with the same number of surviving trees per hectare, basal area growth will increase with increasing dominant stand height. The model developed by Pienaar et al (1990) for *Pinus elliottii* plantations in the Atlantic Coast flatwoods of Georgia takes account of this fact:

$$\ln(G) = \alpha_0 + \alpha_1 \frac{1}{t} + \alpha_2 \ln(H) + \alpha_3(N) \quad ...(6.16)$$

with, H = dominant stand height [m],

t = stand age [years],

G = basal area [m²/ha],

N = stems per ha,

$\alpha_0, \alpha_1, \alpha_2, \alpha_3$ = empirical parameters

The stand basal area is an important density measure, which simultaneously takes into account the average tree size and the number of trees per unit area. Basal area is being used to analyse the relationship between stand density and tree growth (Assmann 1970). Moreover, in combination with the number of trees, basal area can be used to define the type and weight of a thinning (Gadow and Hui 1999, Staupendahl 1999). Modelling basal area growth requires data from plots of varying age and density, re-measured at least once. Models for stand basal area development have been generated by various workers using a differential equation or the path invariant algebraic difference form of a non-linear equation (Pienaar and Shiver 1986, Forss et al 1996, Souter 1986, Hui and Gadow 1993b, Garcia 1994, Rodríguez 1995, Schumacher 1939, Kvist Johannsen 1999).

Pienaar and Shiver (1986) developed a growth function which makes it possible to forecast basal area growth as a function of age, height and stem number:

$$\ln(BA_2) = \ln(BA_1) + \alpha \times \left(\frac{1}{A_2} - \frac{1}{A_1} \right) + \beta \times (\ln N_2 - \ln N_1)$$

$$+ \gamma \times (\ln H_2 - \ln H_1) + \delta \times \left(\frac{\ln H_2}{A_2} - \frac{\ln H_1}{A_1} \right) \quad ...(6.17)$$

where,

BA_1 and BA_2 = basal area at age A_1 and A_2

H_1 and H_2 = top height at age A_1 and A_2

N_1 and N_2 = Number of stems at age A_1 and A_2

α, β, γ and δ = model parameters

Forss et al (1996) used following modified version of this equation for modelling basal area growth in *Acacia mangium* plantations in Indonesia:

$$\ln(BA_2) = \ln(BA_1) + \alpha \times \left(\frac{1}{A_2} - \frac{1}{A_1} \right) + \beta \times (\ln N_2 - \ln N_1) + \gamma \times (\ln H_2 - \ln H_1) \quad ...(6.18)$$

If thinnings are sufficiently heavy and the time period between successive thinnings is small, we can approximate that there will be no change in number of stems per ha at age A_1 and age A_2 and $N_1 = N_2$. In this case equation (6.18) can be simplified to yield equation (6.19):

$$\ln(BA_2) = \ln BA_1 + \alpha \times \left(\frac{1}{A_2} - \frac{1}{A_1} \right) + \beta \times (\ln H_2 - \ln H_1) \quad ...(6.19)$$

Garcia (1994) presented a differential form of an equation which may be modified to correspond to the difference form as follows:

$$BA_2 = BA_1 + \left[\alpha \times BA_1 \times (\gamma - \ln BA_1) \times \frac{1}{\alpha \times A_1 + \delta} \right] \times \Delta t \quad ...(6.20)$$

where, Δt is the interval between measurements.

Kvist Johannsen (1999) developed another differential model for oak in Denmark that can be simplified to yield the following difference form of equation:

$$BA_2 = BA_1 + \left[\alpha \times BA_1^{\beta} \times \exp(-\gamma \times BA_1 - \delta \times H_1) \right] \times \Delta t \quad ...(6.21)$$

Hui and Gadow (1993b) developed the following equation for projecting a known basal area for stands of *Cunninghamia lanceolata* of varying density:

$$BA_2 = BA_1 \times N_2^{1-\alpha \times H_2^{\beta}} \times N_1^{\alpha \times H_1^{\beta} - 1} \times \left(\frac{H_2}{H_1} \right)^{\gamma} \quad ...(6.22)$$

Applying the condition when $N_1 = N_2$, this equation will be simplified as:

$$BA_2 = BA_1 \times \left(\frac{H_2}{H_1} \right)^{\gamma} \quad ...(6.23)$$

The differential form of the model proposed by Rodríguez (1995) for *Pinus pinaster* in Spain can be expressed in the difference form as follows:

$$BA_2 = BA_1 + (\alpha \times BA_1^{\beta} \times A_1^{\gamma}) \times \Delta t \quad ...(6.24)$$

Schumacher (1939) proposed following age dependent basal area model which was later used by Schumacher and Collie (1960), Clutter (1963) and Sullivan and Clutter (1972):

$$\ln(BA_2) = \alpha + (\ln BA_1 - \alpha) \times \left(\frac{A_1}{A_2} \right) \quad ...(6.25)$$

Souter (1986) presented the following model based on Schumacher's equation:

$$\ln(BA_2) = \left(\frac{A_1}{A_2}\right) \times \ln BA_1 + \alpha \times \left(1 - \frac{A_1}{A_2}\right) + \beta \times \left(\ln N_2 - \left(\frac{A_1}{A_2}\right) \times \ln N_1\right) \quad ...(6.26)$$

Applying these equations, basal area growth models have been developed for *E. camaldulensins* (Tewari 2007c), *D. sissoo* (Tewari and Gadow 2008) and *T. undulata* (Tewari 2010) plantations in IGNP area of Rajasthan and *E.* hybrid (Tewari and Singh 2008) and *A. nilotica* (Tewari 2008) plantations in Gujarat. The models developed are:

E. camaldulensis

$$\ln(BA_2) = \ln(BA_1) - 3.0211 \times \left(\frac{1}{A_2} - \frac{1}{A_1}\right) + 0.5055 \times (\ln N_2 - \ln N_1) + 0.66626 \times (\ln H_2 - \ln H_1)$$

D. sissoo

$$\ln(BA_2) = \ln(BA_1) + 2.3593 \times \left(\frac{1}{A_2} - \frac{1}{A_1}\right) + 0.7704 \times (\ln N_2 - \ln N_1) + 0.8095 \times (\ln H_2 - \ln H_1) - 2.1481 \times \left(\frac{\ln H_2}{A_2} - \frac{\ln H_1}{A_1}\right)$$

T. undulata

$$BA_2 = BA_1 \times N_2^{1-0.19396 \times H_2^{0.56317}} \times N_1^{0.19396 \times H_1^{0.56317} - 1} \times \left(\frac{H_2}{H_1}\right)^{3.94394}$$

E. hybrid

$$\ln(BA_2) = \ln(BA_1) - 0.9861 \times \left(\frac{1}{A_2} - \frac{1}{A_1}\right) + 0.4344 \times (\ln N_2 - \ln N_1) + 1.062 \times (\ln H_2 - \ln H_1) + 0.2639 \times \left(\frac{\ln H_2}{A_2} - \frac{\ln H_1}{A_1}\right)$$

A. nilotica

$$\ln(BA_2) = \ln(BA_1) - 1.6416 \times \left(\frac{1}{A_2} - \frac{1}{A_1}\right) + 0.5636 \times (\ln N_2 - \ln N_1) + 0.8758 \times (\ln H_2 - \ln H_1)$$

where,

BA_1 and BA_2 = basal area (m^2) at age A_1 and A_2

H_1 and H_2 = top height at age A_1 and A_2

N_1 and N_2 = Number of stems at age A_1 and A_2

Tewari and Gadow (2005) also used equations (6.17, 6.18, 6.19, 6.22, 6.23, 6.25, 6.26) to model basal area development in *A. indica* plantations in Gujarat. The the parameter estimates and fit statistics obtained are given in Table 6.2.

Table 6.2: Estimated parameters and summary statistics obtained for the basal area models

Model no	**α**	**β**	**γ**	**δ**	**R2**	**SSE***
6.17	-3.9525	0.6103	1.3175	-0.3154	0.949	8.938
6.18	-4.4876	0.5987	1.3055		0.949	8.941
6.19	-4.4901	0.9644			0.942	10.207
6.22	0.9954	-0.7232	0.6581		0.925	13.001
6.23			1.7616		0.901	17.251
6.25	3.1680				0.956	7.635
6.26	3.2987	-0.0185			0.956	7.626

*SSE=sum of squared errors

4. SIMULTANEOUS REGRESSION

Individual tree forest growth and yield models usually employ a set of equations to describe stand development over time. It is not unusual in the forestry literature to treat the same variable as dependent in one equation and independent in another. When a variable is used as both an endogenous (dependent) variable on the left-hand side of one equation and on the right-hand side of another equation, this renders the system of equation simultaneous (Goldberger 1964). A system of equations can also be simultaneous if they share a common error structure (nonzero covariances), even if there are no endogenous variables on the right-hand side (Zellner 1962). Depending on variable interrelationships and model structure, simultaneous estimation of the parameters of the models may be necessary in order to provide estimates that are consistent and efficient (Hasenauer et al 1998).

It is customary in the single-tree growth modelling literature to estimate the parameters in the growth equations separately with ordinary least squares (OLS) rather than simultaneously with some form of generalised least squares or maximum likelihood (Gregoire 1987). One advantage of simultaneous regression procedures is that the joint covariance matrix is estimated without implicitly restricting the between-equation covariances to be zero. Simultaneous regression techniques will lead to more efficient estimators if either (1) endogenous variables appear on the right-hand side, or (2) the contemporaneous correlations among the stochastic error terms are nonzero, or both. A gain in efficiency will increase the precision of the resulting model predictions. In general, the gain in efficiency is higher when the errors among different equations are highly correlated (Judge et al 1988).

If the inter-correlation between the errors of different equations is the only common influence among the equations, the Zellner's (1962) seemingly unrelated regression (SUR) procedure would be appropriate because the equations are related through contemporaneous correlations in the variance-covariance matrix. However, whenever a response or endogenous variable from one model appears as a predictor variable in another model, both the OLS and SUR estimator will be biased and inconsistent due to this stochastic regressor (Judge et al 1988).

Although Theil's (1953) two-stage least squares (2SLS) can be used with stochastic regressors to provide consistent estimates, an even more efficient estimator was invented by Zellner and Theil (1962): three-stage least square (3SLS). By combining Theil's 2SLS procedure with Zellner's SUR procedure, the resulting 3SLS estimators are consistent and asymptotically more efficient than 2SLS. Zellner and Theil's 3SLS regression procedure is a form of feasible generalised least squares (Judge et al 1980).

The basic idea of 2SLS (Johnston 1984) is to replace each right-hand side (RHS) endogenous variable with an estimate obtained by regressing the endogenous RHS variable on all exogenous (fixed) variables in the model (stage 1). This clever procedure replaces a stochastic predictor with a desirable instrumental variable that is contemporaneously uncorrelated with the RHS endogenous variable (Goldberger 1964). In stage 2, least squares regression is used with this instrumental variable along with the exogenous variable to estimate the parameters in a given model. In 3SLS, the variance-covariance matrix is estimated from the residual covariance matrix of the 2SLS procedure, and then the entire system is estimated simultaneously using generalised least squares (stage 3). This retains the consistency property of 2SLS while increasing the asymptotic efficiency of the 3SLS estimator.

Tewari (2007c) used simultaneous regression to develop stem number and basal area prediction models for *E. camaldulensis* plantations in IGNP area of Rajasthan. The nonlinear regression technique (NLS) was used to pick the best model for stem number and best model for basal area. Equation (6.9) was found better for stem number while equation (6.18) performed best for basal area prediction model. Then, these two final models were fitted simultaneously using two-stage least squares (2SLS). Table 6.3 gives the estimated coefficients and standard errors between nonlinear and simultaneous regression techniques by predictor variable. In basal area model (Eq. (6.18)), ln(N) is used as an endogenous predictor variable on the right-hand side. Because of this stochastic predictor, simultaneous regression procedure leads to different estimates for the basal area model (Table 6.3). The NLS standard errors are slightly less compared to the 2SLS standard errors. The NLS and 2SLS results are same for the stem number model because no endogenous variable is used as a predictor variable on the right-hand side.

Table 6.3: Parameter estimates and standard errors for basal area and stem number models by nonlinear (NLS) and two-stage least squares (2SLS) simultaneous regression techniques

Model	Variable	Nonlinear		Simultaneous	
		Estimates	Standard error	Estimates	Standard error
Basal area	α	-3.0211	0.5314	-2.4773	0.5413
	β	0.5055	0.0879	0.3036	0.1406
	γ	0.6626	0.1006	0.6361	0.1157
Stem number	a	0.5027	0.1129	0.5027	0.1129
	b	-3.0526	0.5612	-3.0526	0.5612
	c	21.3766	4.1946	21.3766	4.1946
	d	2.4940	0.6063	2.4940	0.6063

7

Thinning and Individual Tree Growth

1. THINNING

In a managed forest, thinnings usually have a far greater effect on forest development than natural growth. Thus, the development of a managed forest is not only determined by natural tree growth, but also by the weight and type of the prescribed thinning operations. A quantitative description of a thinning is an essential prerequisite for forecasting future stand development.

A thinning reduces the density of a stand and modifies its structure, and thus greatly influences its subsequent medium and long-term development. It may be said that the evolution of managed woodland is mainly the result of a specific sequence of silvicultural *interferences*.

Whereas growth modelling has always attracted much interest among scientists, and continues to be the subject of numerous research projects, *interference modelling* is not very well developed, which is rather surprising in view of (i) the great impact of silvicultural operations on future stand development and (ii) the relevance of the type of a specific thinning operation for estimating product yields (Gadow and Hui 1999).

The modifications of stand structure, though difficult to model in uneven-aged mixed forests, are more easily defined in even-aged regular stands.

1.1 Thinning Weight

A measure for quantifying thinning weight at the stand level, which is sometimes used in conjunction with a yield table, is the reduction in the degree of stocking (B°) caused by the thinning. Kramer (1990) presents a practical method for estimating thinning yields based on the reduction of B°.

Another popular measure for defining the thinning weight is the change in *relative spacing* first introduced by Hart (1928), and which is also known as the Hart-Becking spacing index or S-percent. Relative spacing is defined by the average distance between the trees, expressed as a proportion of dominant stand height which, if square spacing is assumed, can be written as follows:

$$RS = \frac{\sqrt{10000/N}}{H} \qquad ...(7.1)$$

with RS = relative spacing index

H = dominant stand height $[m]$ and N = stems per ha.

This, when used for developing thinning guides where *RS* is being specified, must be solved for *N*. Thus $N = 10000/(H \cdot RS)^2$.

Relative spacing is sometimes used for constructing thinning ranges which define the maximum and the minimum number of trees for a given stand height.

1.2 Type of Thinning

The future development of a forest is not only influenced by the weight, but also by the type of thinning, which is defined by the selective removal of specific members of the population.

The type of thinning is also reflected by the change of the diameter distribution. This change may be described in different ways. At the stand level, diameter distributions are usually not available. In this case, the SG-ratio may be used:

$$SG = \frac{(N_{\text{thn}}/N_{\text{tot}})}{(G_{\text{thn}}/G_{\text{tot}})} \qquad ...(7.2)$$

with N_{thn}, N_{tot} = removed and total number of stems.
G_{thn}, G_{tot} = removed and total basal area.

1.3 Modelling Thinning

A quantitative description of a thinning is an essential prerequisite for casting future stand development. Furthermore, the product assortments that will become available can be predicted more precisely if one is able to project changes caused by thinning. Method has been developed (Staupendahl and Puumalainen 1999) to model the changes of the dbh-distribution of even-aged pure stands on the basis of arithmetic mean diameter. In this method, more emphasis is on the estimation of the removed trees. The dbh-distribution of the remaining stand can easily be calculated by subtracting the class-wise relative frequencies of the cut trees from those of the stand before the thinning. The dbh-distribution before the thinning is regarded as known or may be estimated applying appropriate methods. The relative reduction in number of trees and a thinning index, describing the type of thinning applied, are used as additional model parameters.

In a managed forest, the vertical structure depends on the type of thinning that is applied. In high thinning only the bigger trees are removed while smaller ones may survive for a long time, resulting in a typical structure with two sub-populations. This type of bimodal diameter distributions may be described using a mixed form of the Weibull function (Hessenmöller and Gadow 2001). The function parameters can be derived using estimates of the maximum and minimum stand diameters, the total stand height as well as the basal area for both layers. The joint parameter for the function can be estimated using the relative proportion of stems per hectare in the lower stand stratum, which depends on the thinning type.

The development of a diameter distribution after a thinning is normally modelled using the parameters of a distribution function, which are estimated as a function of age and stand attributes or various methods of stand table projection. However, a very simple approach was used by Hui and Gadow (1996), using the following equation for describing the cumulative distribution function:

$$F = \frac{1}{1 + \exp(a - bX)} \qquad ...(7.3)$$

The parameters of this function may be estimated directly from the percentile values ($XF = 0.5$ and $XF = 0.9$).

The cumulative diameter distribution is defined by the percentile values Dx (where $Dx = XF = 0.5$ or $Dx = XF = 0.9$) and the change of the two percentile values describes the movement of the diameter distribution. It may be assumed that the ratio of the differences between the logarithms of a percentile value and the quadratic mean diameter at two points in time is constant:

$$\ln Dx_2 - \ln Dg_2 = k(\ln Dx_1 - Dg_1) \qquad ...(7.4)$$

Therefore:

$$Dx_2 = Dg_2 \left(\frac{Dx_1}{Dg_1} \right)^k \qquad ...(7.5)$$

The method is simple and very effective for predicting the movement of the diameter distribution in and even-aged plantation.

1.4 Thinning-Worked Example

Tewari and Gadow (2005) presented a worked example to show the effect of low and high thinning on the basal area development in a neem stand in Gujarat. They considered two types of thinning, low thinning and high thinning, at every 5 years with the first thinning at the age of 5 years. Here, five thinnings are shown as an example between 4 and 28 years of age. The thinning weight is expressed as relative stem number removed and defined as:

$$\text{Thinning weight } rN = \frac{N_{\text{removed}}}{N_{\text{total}}}$$

In both the cases (low as well as high thinning), 25% of the stem number is removed at each thinning. The thinning type is given as:

$$NG = \frac{N_{\text{removed}}}{N_{\text{total}}} \bigg/ \frac{G_{\text{removed}}}{G_{\text{total}}} = \frac{rN}{rG}$$

where N = stem number and G = basal area per hectare.

$NG > 1$ indicates a low thinning and $NG < 1$ a high thinning; $NG = 1$ corresponds to a row or random thinning. In the worked example, they evaluated low thinnings with NG values of 1.2, 1.6 and 2 and high thinnings with NG values of 0.9, 0.8, 0.7, 0.6 and 0.5. Since rN has been fixed as 0.25, one can calculate rG values for each NG value in the low and high thinnings.

For estimating basal area in the stand at every stage, they have used the Souter model (equation (6.26)). The basal area development is presented in Fig. 7.1. Figure 7.1(a) shows the highest basal area values for the Low3 option with NG=2.0. But when considering the cumulative yield at the time of the final harvest at the age of 28 years (the yield available in the stand at this age plus the thinning yields), the Low1 option shows the highest total yield because the bigger-sized trees were removed during the thinnings; the number of removed trees being equal.

Similarly, looking at the high thinnings, Figure 7.1(b) shows higher basal area levels in High1 (with NG=0.9), but the total yield of this option is relatively low due to the fact that smaller trees were removed than in the other High-options. The cumulative yield at the time of final harvest at the age of 28 years is highest in the High4 option. In the case of High5 (NG=0.5), it is slightly less as 50% of basal area is removed at each thinning. The total thinning yield is higher than in the other options but the available stand volume at the final age is very low.

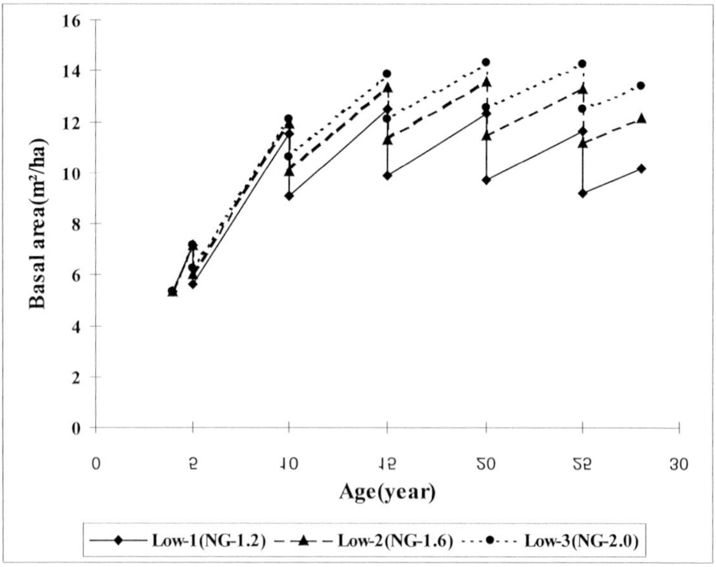

(a)

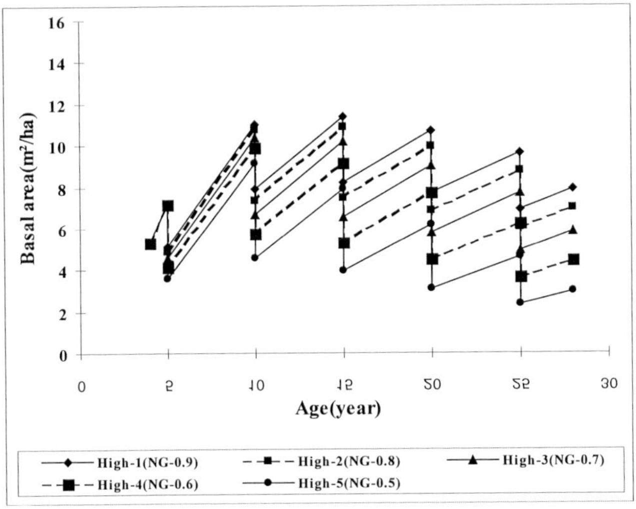

(b)

Fig. 7.1. *Basal area development for neem plantation in Gujarat; (a) low thinning, (b) high thinning.*

1.5 Calculating the Allowable Cut

The different methods of calculating the allowable cut are usually grouped into two kinds, those that restrict the felling area and those that limit the felling volume (Gadow and Bredenkamp 1992). Area control is very popular in plantation forestry, especially when growing conditions and thinning regimes are uniform within the entire forest. The normal felling area is equal to the total forest area divided by the rotation.

The area control method assumes that a regulated forest is desired by the forest department/landowner. When using area control, one would be interested in scheduling for harvest equal areas of land in each time period. By the end of one rotation of using this method, the forest department/landowner should have a forest with equal areas in each age class. In effect, area control stabilizes the area harvested during the conversion period (one rotation, which the forest department/landowner needs to define). One disadvantage of this approach is that the volumes scheduled for harvest may fluctuate widely during the conversion period, particularly when managing an irregular age class forest. This method may be too rigid for the forest department/ landowners that prefer to use natural regeneration.

Volume control is a way to schedule activities to meet specific wood flow requirements of the forest department/landowner each time period. There are a number of methods one can use to determine the annual harvest, from best judgements to a determination from classical methods. While some adjustment period is generally assumed in a forest that has not been regulated, the annual harvest should be recalculated every few years. Volume control provides an even volume harvest over the short-term time horizon, but in the long run, volumes will fluctuate because equal areas are not harvested each year.

Classical forest regulation covers a wide assortment of methods for controlling the cutting volume and for estimating the medium-term yield potential in even-aged forests. These are described below:

(a) Hanzlik formula: $\text{Sustained annual yield} = \left(\frac{V_m}{R}\right) + I$ ...(7.6)

(b) Von Mantel formula: $\text{Sustained annual yield} = \left(\frac{2\ (\text{Growing stock volume})}{\text{Rotation age}}\right)$ ...(7.7)

(c) Austrian formula: $\text{Sustained annual yield} = \left(\frac{G_a - G_r}{a}\right) + I$ ...(7.8)

(d) Hundeshagen formula: $\text{Sustained annual yield} = \left(\frac{G_a}{G_r}\right)(I_r)$ ...(7.9)

(e) Meyer formula: $\text{Annual harvest} = \text{Growth rate}\left(\frac{G_a(1+\text{Growth rate})^n - G_r}{(1+\text{Growth rate})^n - 1}\right)$ ...(7.10)

where, V_m = volume at maturity (m^3/ha), R = rotation (years), G_a = current (actual) growing stock(m^3/ha), G_r = desired future (normal) growing stock (m^3/ha), I = MAI(m^3/ha), I_r = MAI of the future regulated forest(m^3/ha), a = adjustment period (years).

The Hanzlik formula was designed for tracts that had an over-supply of mature timber. The Von Mantel formula is for forests that are nearly regulated, even though the estimate of annual harvest is in the mid-range of those considered. The Austrian formula utilizes an adjustment period to bring a forest to a regulated state (from a volume harvested perspective). The Hundeshagen formula utilizes an estimate of the future growing stock levels. The Meyer formula was designed for forests with some regeneration as well as remnant older forest.

2. FOREST STRUCTURE

Structure is an open-ended theme offering itself to interpretation within many disciplines of the sciences, arts and humanities (Pullan and Bhadeshia 2000). It may develop as a result of a planned design, or through a process of self-organization. Important theoretical concepts relating to biological structures include *self-organisation, structure/property relations* and *pattern recognition.*

The structure of a forest may be defined by the distribution of tree sizes and by the particular relationship between different size variables (*DBH*-height; *DBH*-crown width). Such distributions, however, do not reveal the particular spatial arrangement of these variables. It requires little imagination to realize that a variety of spatial patterns of tree diameters may be possible for any particular diameter frequency distribution. One might refer to a high degree of "size mingling" when large and small trees occur in close vicinity of each other, or to low size mingling when large and small trees are spatially segregated. Such spatial arrangements may be analysed by considering neighbourhood groups of trees. A convenient sampling unit is the *structure-unit,* which may be area-based or neighbourhood-based (Fig. 7.2). An area-based structure unit is defined by the radius of a circle around a sample point, or a sample tree. A neighbourhood-based structure unit may consist of a sample point, or a sample tree, and its n nearest neighbours.

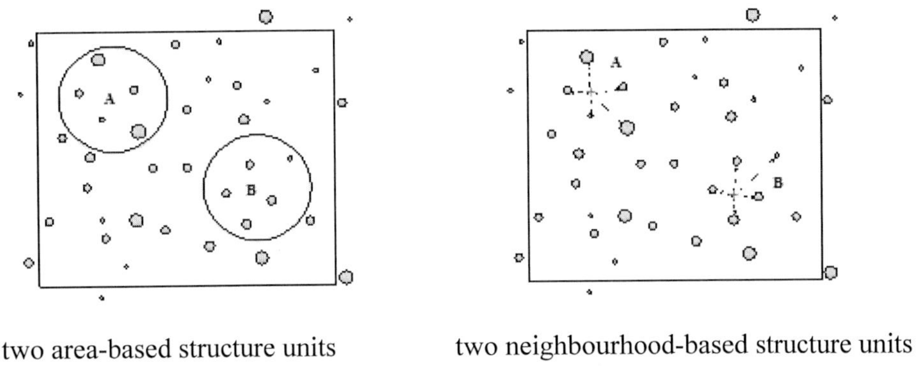

Fig. 7.2. *Spatial mingling of tree species or mingling of tree sizes may be analysed using fixed-area neighbourhood units (FNU) or open-area neighbourhood units (ONU). Special techniques are used to deal with edge effects.*

3. TREE CROWNS

The study of tree morphology is an important aspect of forest growth research and one of the most important morphological elements is the tree crown. Modelling tree crowns is an essential part of forest mensuration and growth modelling because certain crown dimensions are highly correlated with volume, height or diameter growth.

The shape of the tree crown is largely determined by species-specific characteristics and it changes through time as a result of increasing tree age and increasing or decreasing competition (Kramer and Kozlowski 1979, p. 74). It is possible to model tree crowns using simple geometry. Crown lengths are measured using height assessment tools while crown radii are obtained using special devices such as a crown mirror (Van Laar and Akça 1997, p. 76). Photo-digital method may be use (Gaffrey 1996) but it gives inaccurate measurements due to problem associated with the delineation of crown perimeter. Criterion 400 Survey Laser Instrument has also been used for crown radius assessment (Biging and Gill 1997) but the method was found unreliable as the laser beam often passes through the foliage.

Crown window is a new device for assessing crown radii of trees growing in stands of different densities. The device is bias free and produces measurement at low cost with acceptable accuracy (Hussein et al 2000). This instrument is a simple transparent window divided into square sections of equal size. A perforated piece of aluminium (the head stabilizer), mounted on the end of a bar fixed to one side of the frame used to keep the distance between the operator's eye and the crown window constant. The frame is mounted on tripod. Using a suitable horizontal distance, the device is positioned parallel to the stem axis to ensure angular consistency.

4. FOREST DENSITY AND COMPETITON

A basic understanding of the species-specific response to density and competition is a pre-requisite to predicting forest growth and ecosystem dynamics. *Density* is the degree by which available growing space is occupied by a population of trees. *Competition* refers to the restriction or impediment that an individual tree suffers from neighbouring rivals. Trees are sensitive to density and competition because the degree of occupancy of the available growing space by tree crowns and roots affects their access to essential resources and thus their growth and survival. For this reason, one important advantage of tree re-measurement plots is the fact that certain aspects of ecosystem dynamics can be explained by forest density and competition.

Consequently, the literature on density and competition is extensive. Prominent above-ground density measures are basal area (Kramer 1988), relative spacing and stand density index (Reineke 1933), leaf area index (Vose and Allen 1988) and crown competition factor (Krajicek et al 1961). Numerous competition indices, also known as measures of *point density*, have been described in the literature. They include *overlapping zones of influence* (Gerrard 1969), *distance-weighted size ratios* (Hegyi 1974, Lee and Gadow 1997), and *available growing space* (Brown 1965, Hessenmöller 2001). The four methods schematically shown in Fig. 7.3 are representative for numerous similar competition indices that have been published.

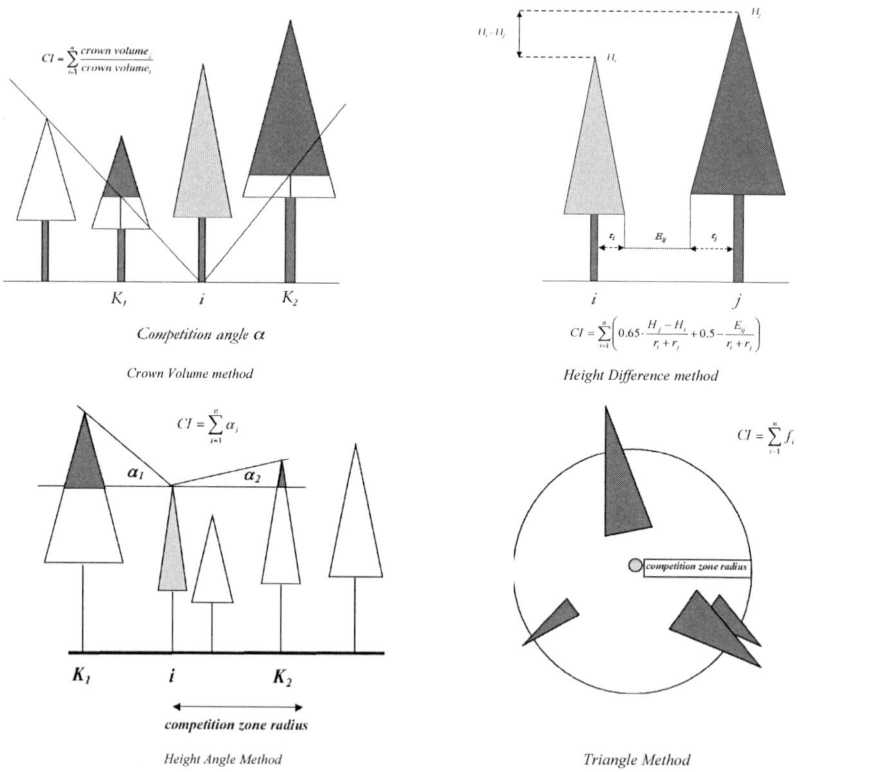

Fig. 7.3. *Four methods for calculating the competition status of a particular reference tree: the crown volume method (Biging and Dobbertin 1992), the height angle method (Rautiainen 1999) the height difference method (Schütz 1989) and the triangle method (Ekö and Ågestam 1994).*

Detailed shading models which simulate the amount of radiation received in three-space are also used, one of the first serious approaches was presented by Courbaud (1995). Light competition modelling involves estimation of the direct and diffuse radiation for any arbitrary point in the plot, as a basis for explaining ecosystem dynamics.

4.1 Competition Index

Inter-tree competition is usually quantified by the competition indices. A competition index is a mathematical formulation derived to represent or describe competition from adjacent trees that could be affecting the growth of any considered tree. There are two major classes of competition index: those which utilise the individual tree location are termed distance-dependent, and those not using locations are termed as distance-independent. One of the commonly used distance-independent competition index in individual tree growth models is BAL index. The basal area of larger trees (BAL) combines individual tree's basal area percentiles with the density measure stand basal area (Wykoff et al 1982). The basal area percentile p_j of a representative tree j is an expression of the relative social rank of a basal area size class within a specified population of trees. The percentile variable considers only a representative tree's competitive position but ignores the density of the stand in which the tree is growing (Schröder and Gadow 1999). Multiplying the percentiles by the stand basal area solves this problem. The BAL index may be given as

$$\mathrm{BAL}_{ij} = BA_i(1 - p_j) \qquad \text{...(7.11)}$$

The resulting BAL index is a linear function of stand density accounting for two different resolution levels simultaneously. Schroder and Gadow (1999) modified this index by adding relative spacing (RS) index and introduced BALMOD as an improvement of BAL index. This alternative approach is to weight the term $(1 - p_j)$ with relative spacing (RSi) of stand *i* as follows:

$$\mathrm{BALMOD}_{ij} = (1 \quad p_j)/\mathrm{RSi} \qquad \text{...(7.12)}$$

5. INDIVIDUAL TREE GROWTH MODELS

The management of mixed stands of tree species is of increasing importance and a crucial factor for sound management of these stands is the knowledge of the growth at individual tree level of different species. Stand level growth models using population mean values are of limited use for mixed un-even aged forests where species and sizes of individual trees are very diverse.

Individual tree growth models can be developed using data either from permanent sample plots observed over long time periods (Monserud and Sterba 1996), from temporary sample plots complemented with increment cores (Wykoff 1990) or from interval plots which represent a compromise between the two (Gadow et al 1998).

Most individual tree growth models can be classified as either polynomial linear models or potential *x* modifier models (Gourlet-Fleury and Houllier 2000). The annual height and dbh growth of a tree, apart from environmental factors, depends on age, tree size, competition and site quality at stand level. Tree age and tree size represent the potential component while competition and site are modifier components, i.e.

$$\text{Growth} = f(\text{age, size, competition, site})$$

Tree growth may be modelled using either basal area or diameter increment. Annual radial increment at breast height may be selected as dependent variable for the diameter growth model.

In some models, tree age has not been included though it influences radial growth. Therefore, age should be taken as an independent variable

$$\text{Age} = f(A) \quad \text{where } A = \text{age}$$

Tree size can be described by a variety of attributes such as dbh, total height and crown length. Crown assessment in the field is expensive. The size variable may be given as:

$$\text{Size } D = f(D) \text{ for a height independent model}$$

$$\text{Size } DH = f(\text{D}, H) \text{ for a height dependent model}$$

A competition component is used as growth modifier. For height independent model, competition index may be written as:

$$CI = f(\text{HgCI}) \text{ with HgCI} = \Sigma(d_j/d_i)/\text{Distt}_{ij}$$

where, d_j = dbh of reference tree (cm), d_i = dbh of competitor tree (cm), Distt_{ij} = distance between reference and competitor tree (m)

For height dependent model, a modified HgCI, as given here, may be used:

$$CI = f(\text{mHgCI}) \text{ with mHgCI} = \Sigma(d_j/d_i)\,(h_j/h_i)/\text{Distt}_{ij}$$

where, h_j = height of reference tree (cm), h_i = height of competitor tree (cm)

For site quality dependent model, site index or topographic site factors may be used:

$$\text{Site} = f(A_0, H_0)$$

where, H_0 = stand dominant height (m), A_0 = stand dominant age (years)

An alternative approach is to directly include specific attributes of the growing site:

$$\text{Site} = f(\text{AI, EL, SL})$$

where, AI = aspect index: AI = 2 × (1-Aspect/180) with values ranging from 0 (south) to 2 (north), EL = elevation (m), SL = slope (degrees)

A system of diameter increment models may be classified according to whether tree height and site quality are included or not.

A general radial increment model may be given as:

$$\Delta r_{t,j} = a_0 A_{t,j}^{a_1}\, D_{t,j}^{a_2}\, H_{t,j}^{a_3}\, e^{bCL_{t,j}0.5} SQ_t \qquad ...(7.13)$$

Diameter growth is directly derived from radial increment. The tree diameter after n years ($D_{t+n,j}$) can be estimated from current diameter ($D_{t,j}$) and the n-year periodical radial increment.

If information about competition is not available, a polynomial linear model for diameter or height increment (Pukkala 1989, Lee et al 2004) may be selected as given below:

$$\Delta Y_{t,j} = a_0 A_{t,j}^{a_1} D_{t,j}^{a_2} H_{t,j}^{a_3} \qquad ...(7.14)$$

where,

$\Delta Y_{t,j}$ = annual dbh or height increment (cm or m, respectively) of tree j ($j = 1 \ldots n$) at time t

$A_{t,j}$ = tree age (years)

$D_{t,j}$ = tree diameter at breast height (cm)

$H_{t,j}$ = tree height (m)

$a_{0,\ldots,3}$ = parameters to be estimated ($a_3 = 0$ in height independent dbh growth model)

Tewari and Gadow (2003) developed individual tree diameter and height growth model for *Prosopis cineraria* stand in Rajasthan using equation (7.14). The parameter estimates and related statistics are presented in Table 7.1.

Table 7.1: Parameter estimates and related statistics for diameter and height increment model for *P. cineraria* in Rajasthan

Parameters	**Diameter growth**	**Height growth**
a_0	0.9504 (0.0039)	0.86325 (0.01419)
a_1	-0.973 (0.002)	-0.924 (0.008)
a_2	0.997 (0.001)	-0.00438 (0.00222)
a_3		0.9975 (0.0044)
RMSE	0.00437	0.00831
Average bias	0.00024	-0.00012
R^2	0.99985	0.996

In brackets standard errors for the parameter estimates are given; RMSE=root mean squared errors

Using similar procedure, individual tree growth models were also developed for an *Ailanthus excelsa* plantation in the hot arid region of Rajasthan state in India (Tewari 2005). The parameter estimates and related statistics are presented in Table 7.2.

Table 7.2: Parameter estimates and related statistics for diameter and height increment model for *A. excelsa* in Rajasthan (equation (7.14))

Parameters	**Diameter growth**	**Height growth**
a_0	0.9648 (0.0029)	0.9273 (0.0069)
a_1	-0.987 (0.001)	-0.956 (0.004)
a_2	1.006 (0.001)	-0.01561 (0.00231)
a_3		0.9761 (0.0034)
RMSE	0.00414	0.00471
Average bias	0.00061	0.00019
R^2	0.99986	0.99864

In brackets standard errors for the parameter estimates are given; RMSE=root mean squared errors

6. DYNAMIC GROWTH MODEL

Reliable growth models are essential for effective long-term planning and decision making. This is especially important in intensively managed forest plantations, where it is necessary to evaluate alternative planting densities, thinning regimes, and rotation lengths (Garcia et al 2011). Empirical models describe and generalize observed stand behaviour. They can be accurate and highly successful in fulfilling many of the forest manager's needs (Burkhart 2008). However, they require extensive long-term experimental data, and extrapolation to unobserved situations can be uncertain. A theory-based model would describe hypothesized behaviour by reasoning from first principles. Theory-influenced semi-empirical models attempt to be compatible with observations, while at the same time use theoretical knowledge and hypotheses to improve performance under conditions for which data are scarce (Mohren and Burkhart 1994).

García et al (2011) gave a new methodological contribution for developing dynamic growth models based on the state-space approach and fitted each transition function separately. In this model, it is assumed that the behaviour of any tree stand evolving in time can be approximated by describing its current state with four state variables: dominant height (H), number of trees per hectare (N), basal area (B), and a measure of site occupancy (Ω), using transition functions to estimate the change of states as a function of the current state of the variables (Garcia 2011, Garcia 2013).

The transition functions are used to predict the growth by updating the state variables, ensuring two natural properties (García 1994): (i) consistency, meaning no change for zero elapsed time; (ii) path-invariance, where the result of projecting the state first from t_0 to t_1, and then from t_1 to t_2, must be the same as that of the one-step projection from t_0 to t_2. Transition functions generated by integration of differential equations (or summation of difference equations when using discrete time) satisfy these conditions and allow computing the future state trajectory. These properties can also be achieved by using techniques for dynamic equation derivation known in forestry as the Algebraic Difference Approach (ADA; Bailey and Clutter 1974) or its generalization (GADA; Cieszewski and Bailey 2000).

Tewari et al (2014b) developed a simple dynamic biologically consistent whole stand growth model for teak stand in Karnataka using state-space approach for modelling rates of change of three state-variables viz., dominant height, stems per hectare and stand basal area. Moreover, the model included a stand volume equation as an output function to estimate this variable at any point in time. Transition functions were fitted separately and simultaneously. A continuous autoregressive error structure was included in the modelling process. For fitting volume equation, generalized method of moments was used to get efficient parameter estimates under heteroscedastic conditions.

6.1 Transition Function for Dominant Height

The two-site-specific parameter equation, derived from the Korf base model and proposed by Tewari et al (2014a), was used as transition function for dominant height. The GADA model allows simultaneous concurrent polymorphism and multiple asymptotes, two characteristics of site equations that are sometimes desirable (Cieszewski 2002). The mathematical expression of this model is

$$H_2 = \exp(X_0)\exp(-(13.6964 + 22.3517/X_0)t_2^{-1.2229}) \text{ with}$$

$$X_0 = \frac{1}{2}t_1^{-1.2229}\left(\begin{array}{l} 13.6964 + t_1^{1.2229}\ln(H_1) + \\ +\sqrt{89.4068t_1^{1.2229} + (13.6964 + t_1^{1.2229}\ln(H_1))^2} \end{array}\right) \quad ...(7.15)$$

where, H_1 and t_1 represent the predictor height (meters) and age (years), and H_2 is the predicted height at age t_2.

6.2 Transition Function for Mortality

Natural mortality or survival is extremely variable and, particularly difficult to predict. The mortality transition function is based on assuming that the rate of change of N relative to dominant height increment depends on the current values of H and N:

$$\frac{dN}{dH} = -a_1 H^{a_2} N^{a_3} \quad ...(7.16)$$

where, N is number of trees ha^{-1}, H is the top height and a_i are parameters to be estimated.

Grouping the terms of equation (7.16) and integrating both sides give the following invariant:

$$\frac{dN}{N^{a_3}} = -a_1 H^{a_2} dH \Rightarrow N^{1-a_3} - \frac{a_1(a_3-1)}{a_2+1}H^{a_2+1} = \text{constant} \quad ...(7.17)$$

Equating the invariant for times 1 and 2 gives the following transition function for number of trees ha^{-1}:

$$N_2 = \left[N_1^{1-a_3} + a_1\frac{a_3-1}{a_2+1}(H_2^{a_2+1} - H_1^{a_2+1})\right]^{1/(1-a_3)} \quad ...(7.18)$$

This predicts N_2 at a height H_2 given any initial values (H_1; N_1). Equation (7.18) was fitted using two different approaches. In the first case, no parameter restrictions were assumed while in the second case, parameter restrictions were obtained by assuming limiting values for the slope of the self-thinning lines based on the average square spacing $S = 100/\sqrt{N}$ (Garcia 2009). The rate of change of average square spacing relative to dominant height increment depends on the initial values of the same variables:

$$\frac{dS}{dH} = \beta^{\gamma}\frac{H^{\gamma-1}}{S^{\alpha-1}} \quad ...(7.19)$$

Integrating equation (7.19) and taking logarithms, the following invariant is obtained:

$$S^{\alpha} = \frac{\alpha}{\gamma}(\beta H)^{\gamma} \Rightarrow \alpha\log(100) - \frac{\alpha}{2}\log(N) = \log(\alpha/\gamma) + \gamma\log(\beta) + \gamma\log(H)$$

$$\Rightarrow \log(\mathrm{N}) = -2\frac{\gamma}{\alpha}\log(H) + \text{constant} \quad ...(7.20)$$

Any limiting value assumed for $2\gamma/\alpha$ in equation (7.20) represents the slope of the self-thinning line assumed for the species. Comparing with equation (7.17), the following relationship is obtained:

$$-2\frac{\gamma}{\alpha} = \frac{a_2+1}{1-a_3} \quad ...(7.21)$$

This relationship could be useful to estimate the parameters of the mortality transition function when the available data is limited, to assure a biologically consistent behaviour (García 2009). Values of 2 and 3 were assumed for the $-2\gamma/\alpha$ ratio, i.e. for the slope of the self-thinning line.

6.3 Transition Function for Basal Area

Instead of predicting basal area (B) directly, we choose to model growth of the product $W = BH$. The rate of change of W can be expressed as the difference between two components: the gross increment and the mortality. In pure and even-aged stands, the gross increment can be written as $b_1\Omega HN^{b_2}$, where Ω is a relative occupancy factor that reduces growth in young or recently thinned stands that do not still fully occupy the site. The mortality as $-k\frac{W}{N}\frac{dN}{dH} = -kW\frac{d\log H}{dH}$, where k represent the mean size of dying trees relative to the mean size of the survivors, considered as constant. Therefore, the general W growth model is:

$$\frac{dW}{dH} = b_1\Omega HN^{b_2} + kW\frac{d\log H}{dH} \quad ...(7.22)$$

A simple closed-form solution of the differential equation (7.22) can be achieved in the special case where b_2 is equal to k:

$$\frac{dW}{dH} = b_1\Omega HN^{k} + kW\frac{d\log H}{dH} \quad ...(7.23)$$

Intuitively, Ω represents resource (e.g. light, nutrient, moisture) interception. In the beginning, occupancy is low in young stands, and gradually increases approaching 1 when canopy closes. Here, it is assumed that the rate of closure initially depends on H in the same way as the gross increment, and decreases to zero as full closure is approached.

$$\frac{d\Omega}{dH} = c_1 H(1 - \Omega) \qquad ...(7.24)$$

Integration gives the invariant:

$$(1 - \Omega)\exp(c_1 H^2/2) = \text{constant} \qquad ...(7.25)$$

With this occupancy model, integration of equation (7.23) gives the transition equation for basal area:

$$B_2 = N_2^k (B_1 H_1 N_1^{-k} + b_1[(H_2^2 - H_1^2)/2 - (\Omega_2 - \Omega_1)/c_1])/H_2 \qquad ...(7.26)$$

with

$$\Omega_2 = 1 - \exp[-c_1(H_2^2 - H_1^2)/2](1 - \Omega_1) \qquad ...(7.27)$$

6.4 Total Stand Volume Equation

Volumes per hectare can be estimated given the state variables using an output function. A total stand volume equation was fitted using as independent variables dominant height and basal area.

$$V = e_1 H^{e_2} B^{e_3} \qquad ...(7.28)$$

This model has been widely used to estimate total stand volume (see, for instance, Husch et al 2003, van Laar and Akça 2007, Weiskittel et al 2011).

The model was fitted using the generalized method of moments (GMM) in the MODEL procedure of SAS/ETS® (SAS Institute Inc. 2008). This method produces efficient parameter estimates under heteroscedastic conditions, without estimating the heteroscedastic error variance (Greene 2012). The main drawback of this method is that error estimates for the predictions cannot be generated without specifying the error structure (Parresol 2001).

6.5 Parameter Estimation and Model Selection

The model was fitted based on projections over pairs of consecutive measurements. It was considered convenient to initialize the state variables at breast height, where $H = 1.37$ and $B = 0$ are known. The number N_b of trees per hectare at this time was estimated by applying an 82.5% survival (since the survival rate observed after one year of planting in these stands was 80-85%) to the density from the spacing adopted at the time of planting which was 2 × 2 m except for one plot in which it was 1.5 × 1.5 m. Following Garcia (2011), the occupancy was initialized at breast height as:

$$\Omega_b = 1 - (1 - \min\{N_b / c_2, 1\})^{c_3} \qquad ...(7.29)$$

where, c_i are parameters to be estimated; c_2 represents a breast-height density at which the stand would be fully closed.

Here, the transition functions for mortality and basal area were fitted simultaneously using the full information maximum likelihood (FIML) in order to obtain estimates that are consistent and efficient and contribute in increased precision of model predictions. Individual tree forest growth and yield models usually employ a set of equations to describe stand development over time. Here, mortality is used as dependent variable in equation (7.18) and as an independent variable in equation (7.26), making the system of equations simultaneous. Simultaneous regression techniques lead to more efficient estimators. A gain in efficiency increases the precision of the resulting model predictions. The gain in efficiency is higher when the errors among different equations are highly correlated (Judge et al 1988).

Age at breast height was estimated using an iterative procedure based on the transition function of dominant height previously described. Those ages were only used to obtain correct consistent estimates of the parameters and their standard errors. The iterative method, proposed by Cieszewski et al (2000), was used for the transition function of dominant height (equation (7.15)). The parameter estimation of the regression equations describing the behaviour of individuals over time often has associated problems of serial correlation. To account for this possible autocorrelation, the error term was modelled using a first-order continuous autoregressive error structure that allows the models to be applied to irregularly spaced, unbalanced data (Gregoire et al 1995, Zimmerman and Núñez-Antón 2001).

6.6 Separate Fitting

Three transition functions for mortality were analyzed and the results are presented in Table 7.3. The first one is based on equation (7.18) without assuming any restrictions for the parameters and the other two are based on assuming values of 2 and 3 for the slope of the self-thinning line according to equation (7.21). The three models explained more than 98% of the total variance, calculated without considering the initialized point at breast height, and the model with a value of 2 for self-thinning line resulted in the best values of the goodness-of-fit statistics.

Table 7.3: Parameter estimates for the mortality transition function (equation (7.18)) and goodness-of-fit statistics calculated without considering the initialized point at breast height

	Parameters			**Goodness-of-fit statistics**	
Model	a_1	a_2	a_3	**RMSE**	**Adj. Model efficiency**
3 parameters	8.43×10^{-10}	1.8792	2.8462	43.1573	0.9845
$\frac{a_2+1}{1-a_3}=3$ *	1.94×10^{-7}	2.0681	2.0227	39.7096	0.9869
$\frac{a_2+1}{1-a_3}=2$ *	1.66×10^{-8}	1.8703	2.4352	38.2869	0.9881

* indicated the parameter value was fixed in the model, therefore, it was not estimated.

For a correct mortality modeling, an adequate combination of short and long intervals between measurements is essential. The time interval between consecutive measurements of the database used in this study was one year and the models fitted showed a good performance for these short periods, within the age interval covered in the sample plots. However, the behavior of the models for long periods projected from the breast age showed high levels of mortality (Fig. 7.4).

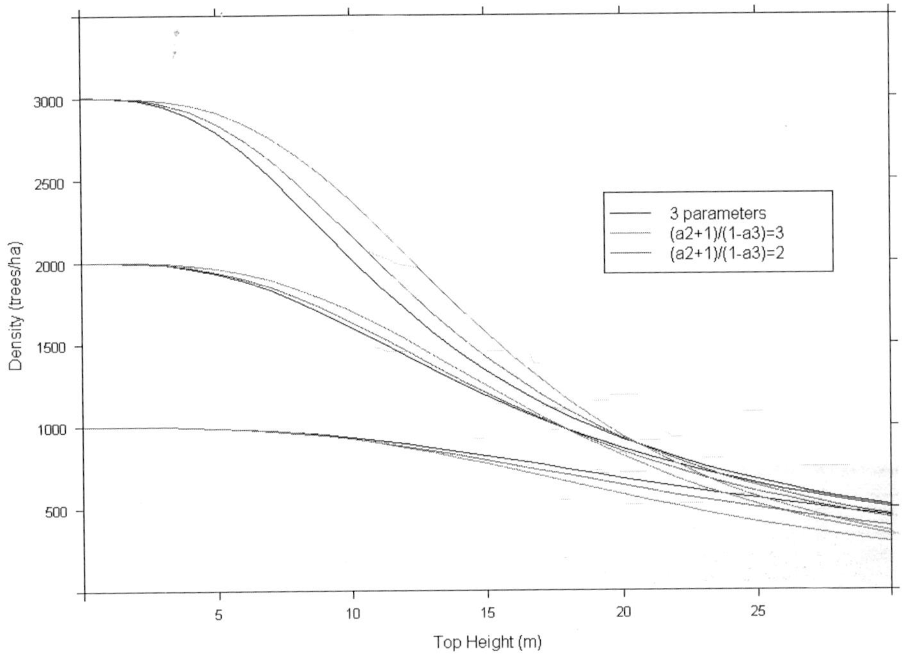

Fig. 7.4. *Data and mortality models. Projections starting from H = 1.37 m and N_1 equal to 1000, 2000 and 3000 trees per hectare*

The results of the fitting of the basal area transition function are shown in Table 7.4. It was not possible to obtain convergence or reasonable estimates with all the parameters. Therefore, as in Garcia et al (2011), c_3 was fixed in equation (7.29) with the value 2.4 and $c2$ in the same equation with the value 10000, implying full canopy closure at breast height with 10000 trees ha^{-1}. It had been shown that these values are not critical (García 2011, García et al 2011).

Table 7.4: Parameter estimates for the basal area transition function (equation (7.26)) and goodness-of-fit-statistics calculated without considering the initialized point at breast height

Parameters			Goodness-of-fit statistics	
k	c_1	b_1	**RMSE**	**Adjusted Model efficiency**
0.2122	0.0369	0.6270	0.5638	0.9939
0.4*	0.0239	0.1844	0.6998	0.9906
0.4*	0.1*	0.1648	0.5636	0.9939
0.4*	0.05*	0.1706	0.6597	0.9917

* indicated the parameter value was fixed in the model, therefore, it was not estimated.

In a first step, the model (equation (7.26)) was fitted without assuming any other restriction for the remaining three parameters (k, b_1 and c_1), however, unrealistically small values for parameters k and c_1 (0.2121 and 0.0369, respectively) were obtained. In a second step, the transition function was fitted assuming a value of 0.4 for parameter k (García 2011), however, the graphical representation indicated that the estimated value of c_1 was unrealistic again (0.0239). Finally, the model was fitted assuming k = 0.4 and two different values for c_1, 0.05 and 0.01. Fig. 7.5 shows occupancy curves for various values of c_1 generated with equation (7.27), starting from 2062 trees per hectare at breast height. Also shown is the relative closure, R, which can be seen as representing the amount of foliage and roots relative to that of a closed stand. Closure and occupancy are related by $\Omega = 1-(1-R)^{c_3}$. Both selected values for c_1, 0.05 and 0.1 seem reasonable, implying nearly closed canopies somewhere between 5 and 10 m top height. Or seen in another way; an equilibrium canopy depth of 5 to 10 m. The best goodness-of-fit statistics was obtained constraining c_1 to 0.05 and this value was used in the simultaneous fitting. The model explained more than 99% of the total variance, calculated without considering the initialized point at breast height.

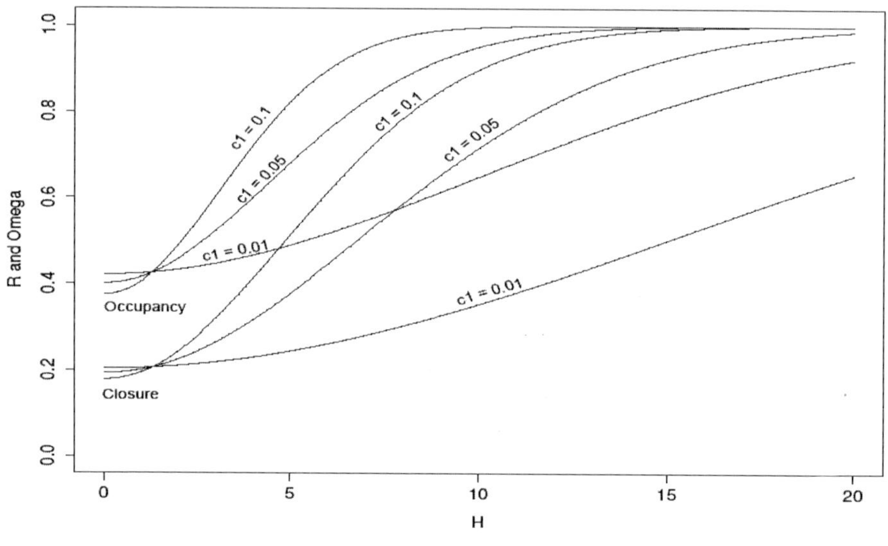

Fig. 7.5. *Closure and occupancy curves*

6.7 Simultaneous Fitting

The results of the simultaneous fitting of both transition functions (equation (7.18) for mortality and equation (7.26) for basal area) are shown in Table 7.5. The goodness-of-fit-statistics were calculated without considering the initialized point at breast height.

Table 7.5: Parameter estimates for the transition functions in mortality and in basal area (equations (7.18) and (7.26), respectively) obtained by fitting the models simultaneously.

Model	Parameter	Estimate	RMSE	Adjusted Model efficiency
Equation (4) $\frac{a_2+1}{1-a_3}=2*$	a_1	1.67×10^{-8}	38.3550	0.9877
	a_2	1.8703		
	a_3	2.4352		
	rho	0.0684		
Equation (12)	k	0.4*	0.5731	0.9937
	c_1	0.1*		
	b_1	0.1663		
	rho	0.0504		

* indicated the parameter value was fixed in the model, therefore, it was not estimated.

Finally, the total stand volume equation was fitted separately using equation (7.28); the model explained more than 97% of the total volume variability with a root mean square error of 13.34 m^3 ha^{-1}. The final expression of the dynamic model was as follows:

$$N_2 = [N_1^{-1.4352} + 0.835 \cdot 10^{-9}(H_2^{2.8703} - H_1^{2.8703})]^{-0.6968} \quad ...(7.30)$$

$$B_2 = N_2^{0.4}(B_1 H_1 N_1^{-0.4} + 0.1663[(H_2^2 - H_1^2)/2 - (\Omega_2 - \Omega_1)/0.1])/H_2 \quad ...(7.31)$$

$$\Omega_2 = 1 - \exp[-0.1[H_2^2 - H_1^2]/2][1 - \Omega_1] \quad ...(7.32)$$

$$\Omega_b = 1 - (1 - \min\{N_b/10000, 1\})^{2.4} \quad ...(7.33)$$

$$V = 0.4628\, H^{0.8063}\, B^{1.1547} \quad ...(7.34)$$

8

Model Evaluation and Validation

Growth models are required to forecast the development of a forest comprising various stands with different species, site attributes and silvicultural regimes and can be used for updating information stored in a data base and for generating data that enable foresters to evaluate alternative silvicultural regimes (Gadow and Hui 1999). Unreliable predictions are at the very least a cause for concern, affecting supply forecasts and profitability calculations. The qualitative and quantitative evaluation of models is therefore a very important part of growth modelling.

1. QUALITATIVE EVALUATION

Qualitative evaluation is done to examine if the model is logically consistent and in agreement with the expected response of a forest to silvicultural treatments. It examines whether model behaviour outside the range of available data is reasonable (Gadow and Hui 1999). An exhaustive qualitative model evaluation is the appropriate measure to make sure that the various process formulations are neither very simple nor inadequate.

2. QUANTITATIVE EVALUATION

A quantitative evaluation of growth models may include characterization of error terms of their magnitude and the distribution of residuals, tests for bias and precision and sensitive analysis to identify those components which have greatest influence on the predictions (Soares et al 1995). Graphical analysis of residuals searching for discrepancies or patterns is an important step in evaluating the fitted models (Gadow and Hui 1999). Residuals should be graphically examined to check for any trend. The linear regression of predicted on observed values should also be done to see the performance of the fitted models. For the purpose, residuals over observed values and observed values over predicted values may be plotted.

The *mean residual MRES*, a measure of average model **bias**, describes the directional magnitude, i.e., the size of expected under- or overestimates. Indices of model **precision** are the *absolute mean residual* (or mean absolute discrepancy), the *root mean square error,* the *model efficiency* and the *variance ratio* (Table 8.1).

The *absolute mean residual* measures the average error associated with a single prediction. The *root mean square error* is based on the residual sum of squares, which gives more weight to the larger discrepancies. The *model efficiency* index is analogous to R^2 and provides a relative measure of performance. The *variance ratio* measures the estimated variance as a proportion of the observed one (Gadow and Hui 1999).

The criteria *mean residual*, *absolute mean residual* and *root mean square error* may be expressed as relative values, which is more revealing when components with different measurement units are compared (Table 8.1).

Table 8.1: Criteria for evaluation model performance

Criterion	Formula	Ideal Value
Mean residual	$MRES = \frac{\sum (y_i - \hat{y}_i)}{n}$	0
Absolute mean residual	$AMRES = \frac{\sum \lvert y_i - \hat{y}_i \rvert}{n}$	0
Root mean square error	$RMSE = \sqrt{\frac{\sum (y_i - \hat{y}_i)^2}{n-1-p}}$	0
Model efficiency	$MEF = \sqrt{\frac{\sum (y_i - \hat{y}_i)^2}{(y_i - \bar{y})^2}}$	0
Variance ratio	$VR = \sqrt{\frac{\sum (\hat{y}_i - \bar{\hat{y}}_i)^2}{(y_i - \bar{y})^2}}$	1

y = observed values; $\hat{y}$ = predicted values; ($y - \hat{y}$) = residuals; p = number of model parameters

3. STATISTICAL TESTS

Formal statistical tests are available for evaluating model performance. Linear regression of predicted on observed values is used to test for bias while Model precision (Freese 1960), a standardized sums of squares criterion, may be used for evaluating precision (Table 8.2).

Table 8.2: Two criteria used in formal statistical tests

Criterion	Formula	Ideal Value
Linear regression	$y = \alpha + \beta \cdot \hat{y}$	$\alpha = 0; \beta = 1$
Model precision	$MPR = \frac{1}{\sigma^2} \sum (y_i - \hat{y}_i)^2$	0

σ^2 *is the variance of y*

A useful test for bias is the simultaneous F-test for slope and zero intercept (Vanclay 1994). The test is based on a linear regression of observations (y_i) against predictions ($\hat{y}_i$) of the form $y = \alpha + \beta \cdot \hat{y}$. For testing the hypothesis that α=0 and β=1, the test criterion F is calculated as follows:

$$F_{2,n} = \frac{(n-2)\left[\left(a\sum y_i + b\sum \hat{y}_i y_i\right) - \left(2\cdot\sum \hat{y}_i y_i - \sum \hat{y}_i^2\right)\right]}{2\cdot\left[\sum y_i^2 - \left(a\sum y_i + b\sum \hat{y}_i y_i\right)\right]}$$

where, n = number of sampling units.

F_α is the threshold value of F (normally at =5% or 1%). The hypothesis is rejected, when $F_{2,n} > F_\alpha$.

4. PSEUDO RESIDUALS

The chi-squared test is generally applied to test goodness of fit of a bivariate distribution function. An alternative and somewhat more versatile tool for assessing the fit, is to use so-called pseudo-residuals (Zucchini and MacDonald 1999). If the observations $x_1, x_2,\ldots,x_n$ are assumed to be independently distributed according to the probability density function f(x) and denote the corresponding probability distribution function by $F(x)$, the pseudo-residuals, r_i, corresponding to the observation x_i is defined as:

$$r_i = \Phi^{-1}\,(F(x_i)),\;\; i = 1, 2, \ldots, n$$

where, Φ denotes the distribution function of the standard normal distribution. The pseudo-residuals are ordered from the smallest to largest.

In diameter (d)-height (h) relationships, in simple regression case only the conditional expectation is a function of d; the conditional variance does not depend on d. Consequently, the ordinary residuals are identically distributed. They can therefore be compared for the purposes of detecting outliers and for assessing the fit of the model. This property does not hold for complex bivariate distributions or for mixture models. For such cases the conditional mean, conditional variance and even the shape of the density function depends on d. Hence ordinary residuals all have different distributions and as such are not comparable. Pseudo-residuals were developed to overcome this problem. They are comparable because they all have same distribution, namely standard normal.

Pseudo-residuals also enable one to assess the fit of a bivariate model in additional detail. We can examine the fit of its projections onto lines other than the horizontal axis or the vertical axis. This enables us to examine the fit from additional points of view, thereby make it possible to detect outliers or lack of fit that might not be less evident if one only considers the usual marginal distributions.

5. CRITERION-BASED PROCEDURES

If there are p potential predictors, then there are 2^p possible models. We fit all models and choose the best one according to some criterion. Some criteria are:

5.1 Akaike's Information Criteria (AIC)

In general,

$$\text{AIC} = n\ln\hat{\sigma}^2 + 2l$$

where, n is the total number of observations used to fit the models; p the number of model parameters; $l = p + 1$, and $\hat{\sigma}^2$ the estimator of the error variance of the model, the value of which is obtained as follows:

$$\hat{\sigma}^2 = \sum_{i=1}^{n} \frac{(y_i - \hat{y}_i)^2}{n}$$

Because AIC is on a relative scale, we usually compute the AIC differences (rather than actual AIC values) over all candidate models in the set. Such differences estimate the relative expected K-L differences between f and $g_i(x \mid \theta)$:

$$\begin{aligned}\Delta_i &= \text{AIC}_i - \text{minAIC}, \\ &= E_{\hat{\theta}}[\hat{I}(f, g_i)] - \min E_{\hat{\theta}}[I(f, g_i)]\end{aligned}$$

These Δ_i values are easy to interpret and allow a quick comparison and ranking of candidate models and are also useful in computing Akaike weights and other quantities of interest (Burnham and Anderson 1998). As a thumb rule, models with $\Delta_i \le 2$ have substantial support and should receive consideration in making inferences. Models having Δ_i of about 4 to 7 have considerably less support, while models with $\Delta_i > 10$ have either essentially no support, and might be omitted from further consideration, or at least those models fail to explain some substantial explainable variation in the data. If observations are not independent but are assumed to be independent then these simple guidelines cannot be expected to hold (Burnham and Anderson 1998).

5.2 Adjusted Coefficient of Determination (R^2_{adj})

Adding a variable to a model can only decrease the Residual Sum of Squares (RSS) and so only increase R^2. Hence, R^2 by itself is not a good criterion because it would always choose the largest possible model.

$$R^2_{adj} = 1 - \frac{n-1}{n-p} \sum_{i=1}^{n} \frac{(y_i - \hat{y}_i)^2}{(y_i - \bar{y}_i)^2} = 1 - [(n-1)/(n-p)](1 - R^2)$$

Adding a predictor will only increase R^2_{adj} if it has some value. Minimizing the standard error for prediction means minimizing $\hat{\sigma}^2$ which in turn means maximizing R^2_{adj}.

5.3 Predicted Residual Sum of Squares (PRESS)

Predicted residual sum of squares (PRESS) is defined as $\sum_i \hat{\varepsilon}^2_{(i)}$ where $\hat{\varepsilon}_i$ are the residuals, calculated without using case i in the fit (Green 1983). The model with the lowest PRESS criterion is then selected. This tends to pick larger models (which may be desirable if prediction is the objective).

5.4 Mallow's C_p Statistic

A good model should predict well so average Mean Squared Error (MSE) of prediction might be good criterion (Faraway 2002):

$$\frac{1}{\sigma^2}\sum_i E(\hat{y}_i - Ey_i)^2$$

which can be estimated by the C_p statistic:

$$C_p = \frac{RSS_p}{\hat{\sigma}^2} + 2p - n$$

where, $\hat{\sigma}^2$ is from the model with all predictors and RSS_p indicates the RSS from a model with p parameters.

(a) C_p is easy to compute.

(b) It is closely related to R^2_{adj} and the AIC.

(c) For the full model $C_p = p$ exactly.

(d) If a p predictor model fits, then $E(RSS_p) = (n-p)\,\sigma^2$ and then $E(C_p) \approx p$. A model with a bad fit will have C_p much bigger than p.

It is usual to plot C_p against p. We desire models with small p and C_p around or less than p.

6. MODEL SELECTION UNCERTAINTY

Akaike's Information Criterion (AIC) and other information theoretic methods can be used to rank the candidate models from best to worst. Often data do not support only one model as clearly best for data analysis. There may be more than one model that is to be regarded as 'useful'. The inability to ferret out a single best model is not a defect of selection criterion; rather, it is an indication that the data are simply inadequate to reach such a strong inference. That is, the data are ambivalent concerning some effect or parameterization or structure.

It is perfectly reasonable that several models would serve nearly equally well in approximating a set of data. Inference must admit that there are sometimes competing models and the data do not support selecting only one. Using the Principle of Parsimony, if several models fit the data equally well, the one with the fewest parameters might be preferred; however, some consideration should be given to the other competing models that are essentially tied as the best approximating model. The issue of competing models is especially relevant in including model selection uncertainty into estimators of precision and model averaging.

One must keep in mind that there is often considerable uncertainty in the selection of a particular model as the 'best' approximating model. The observed data are conceptualized as random variables; their values would be different if another, independent set is available. It is this 'sampling variability' that results in uncertain statistical inference from the particular data set being analyzed. While we would like to make inferences that would be robust to other (hypothetical) data sets, our ability to do so is still quite limited, even with procedures such as AIC, with its cross-validation properties, and with independent and identically distributed sample data. Various computer intensive, re-sampling methods may further improve our assessment of uncertainty of our inferences, but it remains important to understand that proper model selection is accompanied by a substantial amount of uncertainty (Burnham and Anderson 1998).

References

Adame P, Cañellas I, Roig S, and del Rio 2006. Modeling dominant height growth and site index curves for Rebollo oak (*Quercus pyrenaica* Willd.), *Annals of Forest Science* **63** 929–940

Alder D 1980 *Forest volume estimation and yield prediction. Vol. 2 - Yield prediction*, Forestry paper 22/2, Rome: FAO

Alder D 1995 *Growth Modelling for Mixed Tropical Forests*, Oxford: Oxford Forestry Institute

Arabatzis AA and Burkhart HE 1992 An evaluation of sampling methods and model forms estimating height-diameter relationships in loblolly pine plantations, *Forest Science* **38** 192–198

Archibald DJ and Bowling C 1995 Jack pine density management diagram for boreal Ontario, Ontario: Ontario Ministry of Natural Resources, Northeast and Northwest Science and Technology, NEST TN-005, NWST TN-31, 19 pages

Assmann E 1953 Zur Bonitierung süddeutscher Fichtenbestände, *Allgemeine Forst- und Jagd-Zeitung* **10** 61-64

Assman E 1970 *The principles of forest yield study: studies in the organic production, structure, increment, and yield of forest stands*, New York: Pergamon Press

Avery TE and Burkhart HE 1994 *Forest Measurements*, New York: McGraw-Hill

Bailey RL and Clutter JL 1974 Base-age invariant polymorphic site curves, *Forest Science* **20** 155-159

Bailey RL and Dell TR 1973 Quantifying diameter distributions with the Weibull function, *Forest Science* **19** 97-104

Barrio M and Álvarez-González JG 2005 Development of a stand density management diagram for even-aged pedunculate oak stands and its use in designing thinning schedules, *Forestry* **78** 209-216

Bertalanffy Lv 1949 Problems of organic growth, *Nature* **163** 156–158

Bertalanffy Lv 1957 Quantitative laws in metabolism and growth, *Quarterly Review of Biology* **32** 217–231

Bi H 1999 Predicting stem volume to any limit for native tree species in southern New South Wales and Victoria, *New Zealand Journal of Forestry Science* **20** 318-331

Bi H 2000 Trigonometric variable-form taper equations for Australian Eucalypts, *Forest Science* **46** 397-409

Bi H and Hamilton F 1998 Stem volume equations for native tree species in southern New South Wales and Victoria, *Australian Forestry* **61** 275-286

Bigging GS 1985 Improved estimates of site index curves using a varying-parameter model, *Forest Science* **31** 248-259

Bigging GS and Dobbertin M 1995 Evaluation of competion indices in individual tree growth models, *Forest Science* **41**(2) 360-377

Biging GS and Gill SJ 1997 Stochastic models for conifer tree crown profiles, *Forest Science* **43**(1) 25-33

Borders BE, Bailey RL, and Ware KD 1984 Slash pine site index from a polymorphic model by joining (splining) non-polynomial segments with an algebraic difference method, *Forest Science* **30** 411-423

Botkins DB 1993 *Forest dynamics: an ecological model*, Oxford: Oxford University Press

Bravo-Oviedo A, del Río M, and Montero G 2007 Geographic variation and parameter assessment in generalized algebraic difference site index modeling, *Forest Ecology and Management* **247** (1-3) 107-119

Brink C and Gadow Kv 1986 On the use of growth and decay functions for modelling stem profiles, *EDV in Medicine und Biology* **17**(1/2) 20-27

Brown GS 1965 *Point density in stems per acre*, Forestry Research Notes No. 38, New Zealand Forest Service

Bruce D and Schumacher FX 1950 *Forest mensuration*, 3rd Edition, New York: McGraw-Hill

Bruce D, Curtis RO, and Vancoevering C 1968 Development of a system of taper and volume tables for red alder, *Forest Science* **14** 339-350

Burkhart HE 1977 Cubic foot volume of loblolly pine to any merchantable top limit, *Southern Journal of Applied Forestry* **1**(2) 7-9

Burkhart HE 2008 Modelling growth and yield for intensively managed forests, *Journal of Forest Science* **24**(3) 119-124

Burkhart HE and Tomé M 2012 *Modelling Forest Trees and Stands*, Dordrecht: Springer

Burman LE 2010 A call for replication studies, *Public Finance Review* **38**(6) 787–793

Burnham KP and Anderson DR 1998 *Model Selection and Inference. A Practical Information-Theoretic Approach*, New York: Springer-Verlag, New York

Byrne JC and Reed DD 1986 Complex compatible taper and volume estimation systems for red and loblolly pine, *Forest Science* **32**(2) 423-443

Carmean WH and Lenthall DJ 1989 Height-growth and site-index curves for jack pine in north central Ontario, *Canadian Journal of Forest Research* **19** 214-224

Castedo-Dorado F, Crecente-Campo F, Álvarez-Álvarez P, and Barrio-Anta M 2009 Development of a stand density management diagram for radiata pine stands including assessment of stand stability, *Forestry* **82**(1) 1-16

Chapman DG 1961 Statistical problems in population dynamics, 153-168. In *Proceedings of 4th Berkley Symposium on Mathematical Statistics and Probability*, Berkley, University of California Press

Chapman HH 1921 *Forest Mensuration*, New York: John Wiley & sons inc.

Cieszewski CJ 2001 Three methods of deriving advanced dynamic site equations demonstrated on inland Douglas-fir site curves, *Canadian Journal of Forestry Research* **31** 165–173

Cieszewski CJ 2002 Comparing fixed-and variable-base-age site equations having single versus multiple asymptotes, *Forest Science* **48**(1) 7–23

Cieszewski CJ 2003 Developing a well-behaved dynamic site equation using a modified Hossfeld IV function Y3 = (Axm)/ (C + Xm-1), a simplified mixed-model and scant Sub-alpine fir data, *Forest Science* **49** 539–554

Cieszewski CJ 2004 GADA Derivation of dynamic site equations with polymorphism and variable asymptotes from Richards to Weibull and other exponential functions, Athens: Plantation Management Research Cooperative, 10 pages

Cieszewski CJ and Bailey RL 2000 Generalized algebraic difference approach: theory based derivation of dynamic site equations with polymorphism and variable asymptotes, *Forest Science* **46** 116–126

Cieszewski CJ and Nigh G 2002 A dynamic equation for a Sitka Spruce height–age model, *Forest Chronicle* **78**(5) 1–5

Cieszewski CJ and Strub M 2008 Generalized algebraic difference approach derivation of dynamic site equations with polymorphism and variable asymptotes from exponential and logarithmic functions, *Forest Science* **54** 303-315

Cieszewski CJ, Harrison M, and Martin SW 2000 Practical methods for estimating non-biased parameters in self-referencing growth and yield models, University of Georgia, PMRC–TR 2000–7

Clutter JL 1963 Compatible growth and yield models for loblolly pine, *Forest Science* **9** 354-371

Clutter JL 1980 Development of taper functions from variable-top merchantable volume equations, *Forest Science* **26**(1) 117-120

Clutter JL and Jones EP 1980. *Prediction of growth after thinning in old field slash pine plantations,* USDA For. Serv. Res. Paper SE-217

Clutter JL, Fortson JC, Pienaar LV, Brister GH, and Bailey RL 1983 *Timber management: A quantitative approach*, New York: John Wiley and sons.

Creswell JW 2008 *Educational research: Planning, conducting, and evaluating quantitative and qualitative research* (3rd), New Jersey: Prentice Hall

Courbaud B 1995 Modélisation de la croissance en forêt irrégulière, - perspectives pour les pessières irrégulières de montagne, *Revue Forestière Française* **67** 173-182

Curtis RO 1967 Height-diameter and height-diameter-age equations for second-growth Douglas fir, *Forest Science* **13**(4) 365-375

Davis KP 1966 *Forest management: Regulation and Valuation*, Second edition, New York: McGrraw-Hill

Davis LS, Johnson KN, and Davis KP 1987 *Forest management*, New York: McGraw-Hill

Dean JT and Baldwin VC 1993 Using a density management diagram to develop thinning schedules for loblolly pine plantations, New Orleans (LA): USDA Forest Service, Southern Forest Experimental Station (Research Paper no. SO 275)

Dean TJ and Baldwin VC 1996 Crown management and stand density. in *Growing Trees in a Greener World: Industrial Forestry in the 21st century (Proceedings of the 35th LSU Forestry Symposium)* (ed.) M.C. Carter, LA: Louisiana State University Agricultural Center, Louisiana Agricultural Experiment Station, Baton Rouge, 148-159 pages

Demaerschalk JP 1972 Converting volume equations into compatible taper equation, *Forest Science* **18**(3) 241-245

Demaerschalk JP 1973 Integrated systems for the estimation of tree taper and volume, *Canadian Journal of Forest Research* **3**(1) 90-94

Demaerschalk JP and Kozak A 1977 The whole bole system: a conditioned duel-equation system for precise prediction of tree profiles, *Canadian Journal of Forest Research* **7** 488-497

Diéguez-Aranda U, Burkhart HE, and Rodriguez-Soalleiro R 2005 Modeling dominant height growth of radiata pine (*Pinus radiata* D. Don) plantations in north-western Spain, *Forest Ecology and Management* **215** 271–284

Dougherty ER 1990 *Probability and statistics for the engineering, computing and physical sciences*, New Jersey: Prentice-Hall

Drew TJ and Flewelling JW 1979 Stand density management: an alternative approach and its application to Douglas-fir plantations, *Forest Science* **25** 518-532

Ek AR 1971 A formula for white spruce site index curves, Madison: University of Wisconsin, Department of Forestry, Forest Research Note 161

Ekö PM and Ågestam E 1994 A comparison of naturally regenerated and planted Scots pine (*Pinus sylvestris* L.) on fertile sites in southern Sweden, *Forest and Landscape Research* **1** III-126.

Faraway JJ 2002 *Linear models with R, CRC texts in statistical science series v. 63*, London: Chapman & Hall

Fisher RA 1935 *The Design of Experiments*, Edinburgh: Oliver & Boyd

Forss E, Gadow Kv, and Saborowski J 1996 Growth models for unthinned *Acacia mangium* plantations in South Kalimantan, Indonesia, *Journal of Tropical Forest Science* **8**(4) 449-462

Fowler JH and Rennie JC 1988 Merchantable height in lieu of total height in stem profile equations, *Forest Science* **34** 505-511

Freese F 1960 Testing accuracy, *Forest Science* **6** 139-145

Freese F 1961 Relation of plot size to variability: an approximation, *Journal of Forestry* **59** 679

Gadow Kv 1987 *Untersuchungen zur Konstruktion von Wuchsmodellen fur Schnellwuchsige Plantagenbaumarten*, Forstliche Forchungsberichte Munchen No. 7, Munchen: Germany

GadowKv 1990 *Forest Management 476*, The Faculty of Forestry, University of Stellenbosch, Stellenbosch: South Africa

Gadow K v 2005. *Forsteinrichtung*, Göttingen: Universitätsverlag

Gadow Kv and Bredenkamp BV 1992 *Forest Management*, Pretoria: Academica

Gadow Kv and Hui GY 1999 *Modelling Forest Development*, Dordrecht: Kluwer Academic Publishers

Gadow Kv, Rojo A, Álvarez JG, and Rodriguez R 1998 *Ensayos de crecimiento en parcelas permanents, temporales y de intervalo*, Inv. Sistemas y Recursos Forestales, Fuera de Serie no 6

Gaffrey D 1988 Forstamts-und Bestandsindividuelles Sortimentierungs Programm als Mittel zur Plannung, Aushaltung und Simulation, Diplomarbeit, Forstliche Fakultat, Universität Goettingen, Göttingen, Germany

Gaffrey D 1996 Sortenorientiertes Bestandswachstums-Simulationsmodell auf der Basis intraspezifischen, konkurrenzbedingten Einzelbaumwachstums-insbesondere hinsichtlich des Durchmessers-am Beispiel der Douglasie, Dissertation Forschungszentrum Waldökosysteme, Universität Göttingen, Göttingen, Germany

Garcia O 1981 Simplified method-of-moments estimation for the weibull distribution, *New Zealand Journal of Forestry Science* **11**(3) 304-306

Garcia O 1988 Experience with an advanced growth modelling methodology, in *Forest growth modelling and prediction* (eds) AR Ek, SR Shifley and TE Burke, USDA For. Serv. Gen. Tech. Rep. NC-120, 668-675 pages

García O 1994 The state space approach in growth modelling, *Canadian Journal of Forest Research* **24** 1894-1903

García O 2009 A simple and effective forest stand mortality model, International *Journal of Mathematical and Computational Forestry & Natural-Resource Sciences* **1**(1) 1-9

García O 2011 A parsimonious dynamic stand model for interior spruce in British Columbia, *Forest Science* **57**(4) 265-280

García O 2013 Building a dynamic growth model for trembling aspen in western Canada without age data, *Canadian Journal of Forest Research* **43**(3) 256-265

García O, Burkhart HE, and Amateis RL 2011 A biologically-consistent stand growth model for loblolly pine in the Piedmont physiographic region, USA, *Forest Ecology and Management* **262**(11) 2035-2041

Gerrard DJ 1969 *Competition quotient – a new measure of the competition affecting individual forest trees*, Michigan State University Agriculture Experimental Station Research Bulletin No. 20

Goelz JCG and Burk TE 1992 Development of a well-behaved site index equation: jack pine in north central Ontario, *Canadian Journal of Forest Research* **22** 776-784

Goelz JCG and Burk TE 1996 Measurement error causes bias in site index equations, *Canadian Journal of Forest Research* **26** 1585-1593

Goldberger AS 1964 *Econometric Theory*, New York: John Wiley

Goulding CJ 1972 *Simulation technique for a stochastic model of growth of Douglas-fir*. Ph. D. thesis, University of British Columbia, Vancouver, Canada

Goulding CJ 1979 Validation of growth models used in forest management, *New Zealand Journal of Forestry* **24** 108-124

Gourlet-Fleury S and Houllier F 2000 Modelling diameter increment in a lowland evergreen rain forest in French Guiana, *Forest Ecology and Management* **131** 269-289

Gove JH 2004 Structural stocking guides: a new look at an old friend, *Canadian Journal of Forest Research* **34** 1044-1056

Green EJ 1983 Evaluating the predictive abilities of regressions with PRESS, *Forest Science* **29** 712-714

Greene WH 2012 *Econometric analysis*, 7th ed., Upper Saddle River: Prentice Hall

Gregoire TG 1987 Generalized error structure for forestry yield, *Forest Science* **33** 423-444

Gregoire TG, Schabenberger O, and Barrett JP 1995 Linear modelling of irregularly spaced, unbalanced, longitudinal data from permanent-plot measurements, *Canadian Journal of Forestry Research* **25** 137–156

Gregoire TG, Valentine HT, and Furnival GM 1986 Estimation of bole volume by importance sampling, *Canadian Journal of Forestry Research* **16** 554-557

Gregoire TG, Valentine HT, and Furnival GM 1987 Sampling methods for estimating stem volume and volume increment, *Forest Ecology and Management* **21** 311-323

Grosenbaugh LR 1952 Plotless timber estimates – new, fast, easy, *Journal of Forestry* **50** 32-37

Grosenbaugh LR 1958 Point sampling and line sampling: probability theory, geometric implications, synthesis, Occasional Paper No. 160, USDA Southeastern Forest Experimental Station, Asheville NC

Groves RM, Fowler FJ, Couper MP, Lepkowski JM, Singer E, and Tourangeau R 2004 *Survey Methodology*, New York: John Wiley

Grut M 1977 Equations for calculating height increment and site index of Pinaster pine in the Cape Province, *South African Forestry Journal* **102** 43-50

Hafley WL and Schreuder HT 1977 Statistical distributions for fitting diameter and height data in even-aged stands, *Canadian Journal of Forest Res*earch **7** 481-487

Hansen MM and Hurwitz WN 1943 On the theory of sampling from finite populations, *Annals of Mathematical Statistics* **14** 333-362

Hart HMJ 1928 *Stamtal en Dunning - een Orienteerend Onderzoek naar de Beste Plantwijdte en Dunningswijze voor den Djati*, Wageningen: H. Venman & Zonen

Hasenauer H, Monserud RA, and Gregoire TG 1998 Using simultaneous regression techniques with individual-tree growth models, *Forest Science* **44** 87-93

Hegyi F 1974 A simulation model for managing jackpine stands, in *Growth models for Tree and Stand Simulation* (ed.) J. Fries, Stockholm: Royal College of Forestry, 74-90 pages

Hessenmöller D 2001 *Modelle zur Wachstums- und Durchforstungssimulation im Göttinger Kalkbuchenwald*, Dissertation, Fakultät für Forstwissenschaften und Waldökologie, Universität Göttingen, Berlin: Logos Verlag

Hessenmöller D and von Gadow K 2001 Beschreibung der Durchmesserverteilung von Buchenbeständen mit Hilfe der bimodalen WEIBULLfunktion, *Allgemeine Forst-und Jagd-Zeitung* **172**(3) 46-50

Honer TG 1964 The use of height and squared diameter ratios for the estimation of cubic foot volume, *Forest Chronicle* **40** 324-331

Hossfeld JW 1882 *Mathematik für Forstmänner, Ökonomen und Cameralisten*, Gotha, 4. Bd., S. 310

Hradetzky J 1999 Höhenermittlung bei Betriebsinventuren in Baden-Württemberg, *Centralblatt für des gesamte Forstwesen* **116**(1/2) 119–128

http://www.for.gov.bc.ca/hre/gymodels/GY-model

Huang S 1999 Development of compatible height and site index models for young and mature stands within an ecosystem-based management framework, in *Empirical and process-based models for forest tree and stand growth simulation* (eds) A Amaro and M Tome, Lisboa: Portugal, 61-98 pages

Hui GY and Gadow Kv 1993a Zur Entwicklung von Einheitshöhenkurven am Beispiel der Baumart *Cunninghamia lanceolata*, *Allgemeine Forst-und Jagd-Zeitung* **164**(12) 218-220

Hui GY and Gadow Kv 1993b Zur Modellierung der Bestandesdrungflächenentwicklung dargestellt am Beispiel der Baumart bei *Cunninghamia lanceolata*, *Allgemeine Forst-und Jagd-Zeitung* **164** 144-149

Hui GY and Gadow Kv 1996 Ein neuer Ansatz zur Modellierung von Durchmesserverteilungen, *Centralblatt für des gesamte Forstwesen* **113**(¾) 101-113

Husch B 1983 *Forest Mensuration.* 3rd edition, New York: John Wiley and Sons

Husch B, Beers TW, and Kershaw JA 2003 *Forest mensuration.* 4th edition, New York: John Wiley and Sons

Hussein KA, Albert M, and Gadow Kv 2000 The crown window-a simple device for measuring tree crowns, *Forstwissenschaftliches Centralblatt* **119** 43-50

Jack SB and Long JN 1996 Linkages between silviculture and ecology: analysis of density management diagrams, *Forest Ecology and Management* **86** 205-220

Jain RC, Tripathi SP, Kumar VSK, and Kumar S 1996 Volume tables *for Acacia tortilis* plantations based on data collected from KJD Abadi plantations of Khajuwala Range, Chhatisgarh Division, IGNP area, Rajasthan, *Indian Forester* **122** 316-322

Jain RC, Tripathi SP, Singh M, and Kumar VSK 1998 Volume tables for *Azadirachta indica* for Gujarat region, *Indian Forester* **124** 123-133

Johnson NL 1949a Systems of frequency curves generated by methods of translation, *Biometrika* **36** 149-176

Johnson NL 1949b Bivariate distributions based on simple translation systems, *Biometrika* **36** 297-304

Johnston 1984 Econometric methods, New York: McGraw-Hill International

Judge GG, Griffiths WE, Hill RC, and Lee TC 1980 *The theory and practice of econometrics*, New York: John Wiley

Judge GG, Hill RC, Griffiths WE, Lütkepohl H, and Lee TC 1988 *Introduction to the theory and practice of econometrics*. 2nd Edition, New York: John Wiley

Kahn M 1994 *Modellierung der Höhenentwicklung ausgewählter Baumarten in Abhängigkeit vom Standort*, Forstliche Forschungsberichte München, No. 141

Kennel R 1972 *Die Buchendurchforstungsversuche in Bayern von 1870 bis 1970*, Forstliche Forschungsberichte München, No. 7

Kimminis JP 1988 Community organization: method of study and prediction of the productivity and yield of forest ecosystems, *Canadian Journal of Botany* **66** 2654-2672

Kitamura M 1964 Theoretical studies on the estimation of stand volume through the sum of Deckpunkthohen, *Agricultural Science* **4** 365-403

Kiviste AK, Álvarez González JG, Rojo A, and Ruiz AD 2002 *Funciones de crecimiento de aplicación en el ámbito forestal*, Monografía INIA: Forestal No. 4. Ministerio de Ciencia y Tecnología. Instituto Nacional de Investigación y Tecnología Agraria y Alimentaria (INIA), Madrid.

Kleinn C 2007 *Lecture Notes for the Teaching Module Forest Inventory*, Department of Forest Inventory and Remote Sensing, Faculty of Forest Science and Forest Ecology, Georg-August-Universität, Göttingen

Knoebel BR and Burkhart HE 1991 A Bivariate Distribution Approach to Modeling Forest Diameter Distributions at Two Points in Time, *Biometrics* **47**(1) 241-253

Korf V 1939 Příspěvek k matematické definici vzrůstového zákona lesních porostů, *Lesnická práce* **18** 339-356

Kozak A, Munro DD, and Smith JHG 1969 Taper functions and their application in forest inventory, *Forest chronicle* **45** 278-283

Krajicek JE, Brinkman KA, and Gingrich SF 1961 Crown competition. A measure of density, *Forest Science* **7** 35-42

Kramer H 1984 *Grundlagen zur Forstlischen Ertragskunde*, Goettingen: Cuvillier Verlag

Kramer H 1988 *Waldwachstumslehre*, Hamburg und Berlin: Verlag Paul Parey

Kramer H 1990 *Nutzungsplanung in der Forsteinrichtung*, 2nd ed., Frankfurt/Main: J.D. Sauerländer´s Verlag

Kramer H and Akça A 1987 *Leitfadebn zur Waldmeßlehre*, Frankfurt: Sauerländer's Verlag

Kramer H and Akça A 1995 *Leitfaden zur Waldmeßlehre*, Frankfurt/Main: J.D. Sauerländer's Verlag

Kramer PJ and Kozlowski TT 1979 *Physiology of woody plants*, New York: Academic Press

Krumland B and Eng H 2005 Site index systems for major young-growth forest and woodland species in northern California, California Department of Forestry and Fire Protection, California Forest Report No. 4

Kumar BM, Long JN, and Kumar P 1995 A density management diagram for teak plantations of Kerala in peninsular India, *Forest Ecology and Management* **74** 125-131

Kvist Johannsen V 1999 *A growth model for oak in Denmark*, Ph.D. Dissertation, Danish Forest and Landscape Research Institute, Hørsholm, Denmark

Landberg JJ 1986 *Physiological ecology of forest production*, London: Academic Press

Lee WK and Gadow Kv 1997 Iterative bestimmung der konkurrenzbäume in *Pinus densiflora* beständen, *Allgemeine Forst-und Jagd-Zeitung* **168**(3/4) 41-44

Lee WK, Gadow Kv, Chung DJ, and Lee JL 2004 Individual tree dbh growth in *Pinus densiflora* and *Quercus variabilis* mixed stands in central Korea, *Ecological Modelling* **176** 187-200

Lilieholm RJ, Long JN, and Patla S 1994 Assessment of goshawk nest area habitat using stand density index, *Studies in Avian Biology* **16** 18-23

Loetsch F, Zohrer F, and Haller KE 1973 *Forest Inventory*, Vol. II, Munchen: BLV Verlagsgesellschaft

Lohr S 1999 *Sampling Design and Analysis*, Pacific Grove: Brooks/Cole Publishing Company

Long JN 1985 A practical approach to density management, *Forestry Chronicle* **23** 23-26

Long JN, McCarter JB, and Jack SB 1988 A modified density management diagram for coastal Douglas fir, *Western Journal of Applied Forestry* **3** 88-89

Long JN and Shaw JD 2005 A density management diagram for even-aged ponderosa pine stands, *Western Journal of Applied Forestry* **20** 205-215

Longchang L, Wenkang H, and Guoquing W 1991 Study on the method of constructing variable-density yield table, *Journal of Northeast Forestry University* **2**(2) 95-99

Luis JFS and Fonseca TF 2004 The allometric model in the stand density management of *Pinus pinaster* Ait. in Portugal, *Annals of Forest Science* **61** 807-814

Lundgren AL and Dolid WA 1970 Biological growth functions describe published site index curves for Lake States timber species. USDA Forest Service, North Central Forest Expental Station, Researh Paper NC-36, 8 pages

Lundqvist B 1957 On the height growth in cultivated stands of pine and spruce in Northern Sweden, *Medd. Fran. Statens Skogforsk.* **47**(2) 1-64

Mäkelä A 1992 Process-oriented growth and yield models: Recent advances and future prospects, 85-96. *In* T. Preuhsler [ed], *Research on growth and yield with emphasis on mixed stands. Proceedings of IUFRO Centennial Meeting S4.01 session on 'Mensuration, Growth and Yield'*, Bayerische Forstliche Vesuchs und Forschungsanstalt, Freising, Germany.

Max TA and Burkhart HE 1976 Segmented polynomial regression applied to taper equations, *Forest science* **22** 283-289

McCarter JB and Long JN 1986 A lodgepole pine density management diagram, *Western Journal of Applied Forestry* **1** 6-11

Medawar PB 1940 Growth, growth energy, and ageing of the chicken's heart, *Proceedings of the Royal Society London B. Biological Sciences* **129** 332-355

Mendoza GA and Vanclay JK 2008 Trends in forestry modelling, *CAB Reviews: Perspectives in Agriculture, Veterinary Science, Nutrition and Natural Resources* **3**(10) 8 pages

Meyer HA 1953 *Forest mensuration*, Pennsylvania: Penns Valley Publishers

Meyer WH 1930 Diameter distribution series in even-aged forest stands, *School of Forestry, Yale University Bull*etin **28** 105 pages

Michailoff JL 1943 Zahlenmässiges Verfahren für die Ausführung der Bestandeshöhenkurven, *Forst-wissenschaftliches Centralbatt - Tharandter Forstliches Jahrbuch* **6** 273–279

Mirkovich D 1958 *Normale visinske kriva za chrast kitnak ibukvu VNR srbiji*, Zagreb: Glasnik Sumarskog Fakulteta No. 13.

Mohren GMJ and Burkhart HE 1994 Contrasts between biologically-based process models and management-oriented growth and yield models, *Forest Ecology and Management* **69** 1-5

Monserud RA and Sterba H 1999 Modeling individual tree mortality for Austrian forest species, *Forest Ecology and Management* **113** 109–123

Munro DD 1974 Forest Growth Models - A Prognosis in *Growth Models for Tree and Stand Simulation* (ed) J Fries, Stockholm: Royal College of Forestry of Sweden, Research Note No. 30, 7-21 pages

Nachtsheim CJ 2009 Split-plot Designs: What, Why, and How, *Journal of Quality Technology* **41**(4) 340-360

Nagel J and Kehr IC 1997 Instant Yield Programme (IYP), Paper presented at the XI World Forestry Congress, Antalya, Turkey

Nelder JA 1962 New kinds of systematic designs for spacing experiments, *Biometrics* **18** 283-307

Newnham RM 1988 A modification of the Ek-Payandeh non-linear regression model for site-index curves, *Canadian Journal of Forest Research* **18** 115-120

Newton PF 1997 Stand density management diagrams: review of their development and utility in stand-level management planning, *Forest Ecology and Management* **98** 251-265

Newton PF 2003 Yield prediction errors of a stand density management program for black spruce and consequences for model improvement, *Canadian Journal of Forest Research* **33** 490-49

Newton PF and Weetman GF 1994 Stand density management diagram for managed black spruce stands, *Forestry Chronical* **70** 65-74

Newton PF, Lei Y, and Zhang SY 2005 Stand-level diameter distribution yield model for black spruce plantations, *Forest Ecology and Management* **209** 181-192

Nord-Larsen T 2006 Developing dynamic site index curves for European beech (*Fagus sylvatica* L.) in Denmark, *Forest Science* **52** 173-181

O'Regan WG and Arvanitis LG 1966 Cost-effectiveness in forest sampling, *Forest Science* **12** 406-414

Oliveira AM 1980 *Untersuchungen zur Wuchsdynamik junger Kiefrbestande*, Ph.D. Thesis, Faculty of Forestry, University of Munich, Germany

Oliveira AM 1987 The H/D ratio in maritime pine (*Pinus pinaster*) stands, Paper presented at the IUFRO Conference on "Forest growth modelling and prediction" held at Minneapolis, USA during Aug. 24-28, 1987

Ortega A and Montero G 1988 Evaluación de la calidad de las estaciones forestales. Revisión bibliográfica, *Ecología* **2** 155-184

Osmaston FC 1968 *The management of forests*. London: George Allen & Unwin

Parde J 1961 *Dendrometry*, Nancy: Edictions de l'Ecole Nationale des Eaux et Forests

Parresol BR 2001 Additivity of nonlinear biomass equations, *Canadian Journal of Forest Research* **31** 865-878

Parresol BR and Vissage JS 1998 White Pine Site Index for the Southern Forest Survey, Research Paper SRS-10, US Department of Agriculture, Forest Service, Southern Research Station, Asheville, NC

Páscoa F 1987 *Estructura, crecimento e produção em povomentos de pinheiro bravo. Um modelo de simulação*. Ph.D. Thesis, Universidade Técnica de Lisboa

Patterson DW, Wiant Jr. HV, and Wood GB 1993 Errors in estimating the volumes of butt logs, *Forest Products Journal* **43**(3) 41-44

Payandeh B and Wang Y 1994 Relative accuracy of a new base-age invariant site index model, *Forest Science* **40** 341-347

Perez LD and Kanninen M 2003 Provisional equations for estimating total and merchantable volume for *Tectona grandis* trees in Costa Rica, *Forests, Trees and Livelihoods* **13** 345-359

Philip MS 1994 *Measuring Trees and Forests*, 2nd edition, Wallingford: CAB International

Pienaar LV and Shiver BM 1986 Basal area prediction and projection equations for pine plantations, *Forest Science* **32** 626-633

Pienaar LV, Page H, and Rheney JW 1990 Yield prediction for mechanically site-prepared slash pine plantations, *Southern Journal of Applied Forestry* **14** 104-109

Pienaar LV 1991 PMRC yield prediction system for slash pine plantations in the Atlantic coast flatwoods, PMRC Technical Report, Athens, USA (cited in Van Laar A and Akça A 1997 *Forest Mensuration*, Göttingen, Germany: Cuvillier Verlag)

Pienaar LV, Harison WM and Rheney JW 1991 Yield prediction system for slash pine plantations in the Atlantic coast flatwoods, PMRC technical report, School of Forest Resources, University of Georgia, Athens, Georgia

Piennar LV and Turnull KJ 1973 The Chapman-Richards generalization of von Bertalanffy's growth model for basal area growth and yield in even-aged stands, *Forest Science* **19** 2-22

Prakash R 1986 *Forest Management*, Dehradun: International Book Distributors

Pretzsch H 1995 Perspektiven einer modellorientierten Waldwachstumsforschun, *Forstwissenschaftliches Centralblatt* **114** 188-209

Prodan M 1953 Verteilung des Vorrats gleichaltriger Hochwaldbestände auf Durchmesserstufen, *Allgemeine Forst- und Jagdzeitung* **129** 15-33

Prodan M 1965 *Holzmeßlehre*, Fankfurt a.M.: J.D. Sauerländer's Verlag

Prodan M 1968 *Forest Biometrics*, Oxford: Pergamon Press

Prodan M, Pepers R, Cox F, and Real P 1997 Mensura Forestal, Instituto Interamericano de cooperación para la agricultura, Serie investigación y educación en desarrollo sostenible l, San Jose, Costa Rica

Pukkala T 1989 Predicting diameter growth in even-aged Scots pine stands with a spatial and non-spatial model, *Silva Fennica* **23** 101-116

Pullan W and Bhadeshia H (eds) 2000 *Structure in Science and Art*, Cambridge: Cambridge University Press

Ramirez-Maldonado H, Bailey RL, and Borders BB 1988 Some implications of the algebraic difference form approach for developing growth models, in *Forest growth modelling and prediction* (eds) AR Ek, SR Shifley and TE Burke, USDA For. Serv. Gen. Technical Report NC-120, 731-738 pages

Rautiainen O 1999 *Growth Dynamics and Management of Shorea robusta Forests in Southern Nepal*, Dissertation, University of Joensuu, Finland

Reineke LH 1933 Perfecting a stand density index for even-aged stands, *Journal of Agricultural Resear*ch **46** 627-638

Reynolds MR, Burkhart HE, and Daniels RF 1981 Procedures for statistical validation of stochastic simulation models, *Forest Science* **27** 349-364

Richards FJ 1959 A flexible growth function for empirical use, *Journal of Experimental Bot*any **10**(29) 290-300

Riemer T, Gadow Kv, and Sloboda B 1995 Ein Modell zur Beschribung von Baumschaften, *Allgemeine Forst und Jagdzeitung* **166**(7) 144-147

Raza L and Masood MA 2009 Efficiency of lattice design in relation to randomized complete block design in agricultural field experiments, *Pakistan Journal of Agriculture Research* **22**(3-4) 150-153

Robinson AP and Wood GB 1994 Individual tree volumen estimation: A new look at new systems, *Journal of Forestry* **92**(12) 25-29

Rodriguez Soalleiro R 1995 *Crecimiento y producción de masas forestales regulares de Pinus pinaster Ait. en Galicia. Alternativas selvícolas posibles*, Ph.D. Dissertation, Escuela Técnica Superior de Ingenieros de Montes, Madrid

SAS Institute Inc 2004 *SAS/ETS User's Guide Version 9*, SAS Institute Inc, Cary, NC

SAS Institute Inc 2008 *SAS/ETS® 9.2 User's Guide*, SAS Institute Inc, Cary, NC

Schumacher FX and Collie TS 1960 *Growth and yield of natural stands of the southern pines*. Durham, N.C.: T.S. Collie Inc.

Schreuder HT and Hafley WL 1977 A useful bivariate distribution for describing stand structure of tree heights and diameters, *Biometrics* **33** 471-478

Schreuder HT and Swank WT 1974 Coniferous stands characterized with the Weibull distribution, *Canadian Journal of Forest Research* **4** 518–523

Schreuder HT, Hafley WL, and Bennett FA 1979 Yield Prediction for Unthinned Natural Slash Pine Stands, *Forest Science* **25**(1) 25–30

Schröder J and Gadow Kv 1999 Testing a new competition index for maritime pine in north-western Spain, *Canadian Journal of Forest Research* **29** 280-283

Schröder J and Gonzalez GA 2001 Comparing the performance of generalized diameter-height equations for maritime pine in northwestern Spain, *Forstwissenschaftliches Centralbatt* **120** 18-23

Schumacher FX 1939 A new growth curve and its applications to timber-yield studies, *Journal of Forestry* **37** 819-820

Schütz JP 1989 Der Plenterbetrieb: Unterlage zur Vorlesung Waldbau III. (Waldverjüngung), Fachbereich Waldbau ETH , Zürich, Switzerland

Scott CT and Voorhis NG 1986 Northeastern forest survey site index equations and site productivity classes, *Northern Journal of Applied Forestry* **3** 144-148

Shugart HH 1984 *Theory of Forest Dynamics*, New York: Springer Verlag

Sloboda B 1971 Zur Darstellung von Wachstumsprozessen mit Hilfe von Differential gleichungen erster Ordnung, Mitteillungen der Badenwürttembergischen Forstlichen Versuchs und Forschungsanstalt, Universität Göttingen, Göttingen, Germany

Smaltschinski T 1997 Großregionale Holzaufkommensprognosen und nationale Forstinventuren, Unpublished manuscript, Universität Göttingen, Göttingen, Germany

Smith FW and Long JN 1987 Elk hiding and thermal cover guidelines in the context of lodgepole pine stand density, *Western Journal of Applied Forestry* **2** 6-10

Soares P, Tome M, Skovsgaard JP, and Vanclay JK 1995 Evaluatiing a growth model for forest management using continuous forest inventory data, *Frest Ecology and Managemen*t **71** 251-265

Souter RA 1986 *Dynamic stand structure in thinned stands of naturally regenerated loblolly pine in the Georgia Piedmont*, Ph.D. Thesis, University of Georgia, Athens, GA

Speidel G 1972 *Planung im forstbetrieb*. Hamburg & Berlin: Paul Parey

Spurr SH 1952 *Forest Inventory* New York: John Wiley and sons

Spurr SH and Barnes BV 1980 *Forest Ecology*, New York: Wiley & Sons, New York

Stage AR 1963 A mathematical approach to polymorphic site index curves for Grand-Fir, *Forest Science* **9**(2) 167-180

Stage AR and Rennie JC 1994 Fixed-radius plots or variable-radius plots? *Journal of Forestry* **92**(12) 20-24

Staupendahl K 1999 *Modelling thinnings based on the ratio of relative removal rates. Growth and yield modelling of tree plantations in south and east Africa*, University of Joensuu, Finland

Staupendahl K and Puumalainen J 1999 Modellierung des Einflusses von Durchforstungen auf die Durchmesserverteilung von gleichaltrigen Fichtenreibeständen, *Centralblatt für des gesamte Forstwesen* **116**(4) 249-262

Sterba H 1975 Assmanns Theorie der Grundflächenhaltung und die "Competition-Density-Rule" der Japaner Kira, Ando und Tadaki, *Centralblatt für das gesamte Forstwesen* **92**(1) 46-62

Sterba H 1987 Estimating potential density from thinning experiments and inventory data, *Forest Science* **33** 1022-1034

Sterba H 1995 Forest decline and increasing increments: a simulation study, Estimación de la productividad forestall con curves de sitio de forma y escala variables. Centro de Investigación y Docencias Económicas, E-110

Sullivan AD and Clutter JL 1972 A simultaneous growth and yield model for loblolly pine, *Forest Science* **18** 76-86

Tewari VP 2000 Fitting of statistical distributions to diameter measurements of tree stand, *Annals of Forestry* **8**(2) 274-278

Tewari VP 2004 Stem number development and potential stand density in the unthinned even-aged *Azadirachta indica* plantations in Gujarat, India, *International Journal of Forestry Review* **6**(1) 51-55

Tewari VP 2005 Individual tree growth models for an *Ailanthus excelsa* Roxb. plantation in the hot arid region of India, in *Multipurpose Trees in the Tropics: Management and Improvement Strategies* (eds) VP Tewari and RL Srivastava, Jodhpur: Scientific Publishers, 428-436 pages

Tewari VP 2006 Generalised diameter-height equations for even-aged stands of *Azadirachta indica* in Gujarat State of India, *Indian Forester* **132**(1) 79-90

Tewari VP 2007a Comparing the model forms estimating generalised diameter-height relationships in *Tecomella undulata* plantations in hot arid region of India, *Journal of Forestry Research* **18**(4) 255-260

Tewari VP 2007b Total wood volume equations and their validation for *Tecomella undulata* plantations in hot arid region of India, *Indian Forester* **133**(12) 1648-1658

Tewari VP 2007c Stand density and basal area prediction of unthinned irrigated plantations of *Eucalyptus camaldulensis* in the hot desert of India, *Bioresource Technology* **98**(5) 1106-1114

Tewari VP 2008 Predicting potential density and basal area development in social forestry plantations of *Acacia nilotica* in Gujarat state of India, *Journal of Sustainable Forestry* **26**(4) 284-300

Tewari VP 2010 Limiting stand density and basal area projection models for even-aged *Tecomella undulata* plantations in a hot arid region of India, *Journal of Forestry Research* **21**(1) 13-18

Tewari VP 2012 Variable density yield model for irrigated plantations of *Dalbergia sissoo* grown under hot arid conditions in India, *Journal of Forest Science* **28**(4) 205-211

Tewari VP and Gadow Kv 1997 Fitting a bivariate distribution to diameter-height data of forest trees, *Indian Forester* **123**(9) 815-820

Tewari VP and Gadow Kv 1999 Modelling the relationship between tree diameters and heights using S_{BB} distribution, *Forest Ecology and Management* **119** 171-176

Tewari VP and Gadow Kv 2003 Preliminary growth models for *Prosopis cineraria* (L.) Druce plantations in the hot arid region of India, *Forests, Trees and Livelihoods* **13**(4) 361-373

Tewari VP and Gadow Kv 2005 Basal area growth of even-aged *Azadirachta indica* stands in the Gujarat State of India, *Journal of Tropical Forest Science* **17**(3) 349-361

Tewari VP and Gadow Kv 2008 Modelling potential density limiting survival, stand density and basal area growth for pure even-aged *Dalbergia sissoo* stands in a hot arid region of India, *Forests, Trees and Livelihoods* **18**(2) 133-150

Tewari VP and Kumar VSK 2001 Construction and validation of tree volume functions for *Dalbergia sissoo* grown under irrigated conditions in the hot desert of India, *Journal of Tropical Forest Science* **13**(3) 503-511

Tewari VP and Kumar VSK 2002 Development of top height model and site index curves for *Azadirachta indica* A. juss, *Forest Ecology and Management* 1**65**(1-3) 67-73

Tewari VP and Kumar VSK 2003 Volume equations and their validation for irrigated plantations of *Eucalyptus camaldulensis* in the hot desert of India, *Journal of Tropical Forest Science* **15**(1) 136-146

Tewari VP and Kumar VSK 2005 Growth and yield functions for *Dalbergia sissoo* plantations in the hot desert of India grown under irrigated conditions, *Journal of Tropical Forest Science* **17**(1) 87-103

Tewari VP and Singh B 2005 Comparison of Bruce's formula and other methods for log volume estimation, *Indian Forester* **131**(7) 417-423

Tewari VP and Singh B 2006a Provisional equations for estimating total and merchantable wood volume of *Acacia nilotica* trees in Gujarat State of India, *Forests, Trees and Livelihoods* **16**(3) 277-288

Tewari VP and Singh B 2006b Total and merchantable wood volume equations for *Eucalyptus hybrid* trees in Gujarat State, India, *Arid Land Research and Management* **20**(2) 147-159

Tewari VP and Singh B 2008 Potential density and basal area prediction equations for unthinned *Eucalyptus* hybrid plantations in the Gujarat state of India, *Bioresource Technology* **99**(6) 1442-1449

Tewari VP and Singh B 2009 Site index model for *Tecomella undulata* plantations in a hot arid region of India, *Journal of Arid Environments* **73**(4-5) 590-593

Tewari VP and Singh B 2018 Total wood volume equations for *Tectona grandis* linn f. Stands in Gujarat, India, *Journal of Forest and Environmental Science* **34**(4) 313-320

Tewari VP and Álvarez-González JG 2014 Development of a stand density management diagram for teak stands in Karnataka (India), *Journal of Forest Science* **30**(3) 259-266

Tewari VP, Álvarez-González JG, and Gadow Kv 2014a Dynamic base-age invariant site index models for *Tectona grandis* in peninsular India, *Southern Forests: a Journal of Forest Science* **76**(1) 21-27

Tewari VP, Álvarez-González JG, and García O 2014b Developing a simple dynamic growth model for teak plantations in India, *Forest Ecosystems* 1: 9

Tewari VP, Corral Rivas JJ, Vilcko F, and Gadow Kv 2007 Height-growth and site index equations for social forestry plantations of *Acacia nilotica* and *Eucalyptus hybrid* in Gujarat State of India, *Forests, Trees and Livelihoods* **17**(2) 125-140

Tewari VP, Mariswamy KM, and Arun Kumar AN 2013 Total and merchantable volume equations for *Tectona grandis* Linn. F. plantations in Karnataka, India, *Journal of Sustainable Forestry* **32**(3) 213-229

Tewari VP, Verma A, and Kumar VSK 2002 Growth and yield functions for irrigated plantations of *Eucalyptus camaldulensis* in hot desert of India, *Bioresource Technology* **85**(2) 137-146

Theil H 1953 *Repeated least-squares applied to complete equation systems*, The Central Planning Bureau, The Hague, The Netherlands.

Trincado G, Gadow Kv, and Tewari VP 1996 Comparison of three stem profile equations for *Quercus robur L., South African Forestry Journal* **177** 23-30

Trincado GV, Kiviste A, and Gadow Kv 2003 Preliminary site index models for native Roble (*Nothofagus obliqua*) and Rauli (*N. alpina*) in Chile, *New Zealand Journal of Forest Science* **32**(3) 322–333

Valbuena P, Del Peso C, and Bravo F 2008 Stand Density Management Diagrams for two Mediterranean pine species in Eastern Spain, *Investigación Agraria: Sistemas y Recursos Forestales* **17**(2) 97-104

Van Deusen PC 1987 Combining taper functions and critical height sampling for unbiased stand volume estimation, *Canadian Journal of Forestry Research* **17**(1) 416-420

Van Deusen PC and Lynch TB 1987 Efficient unbiased tree-volume estimation, *Forest Science* **33** 583-590

Van Laar A and Akça A 1997 *Forest Mensuration*, Göttingen, Germany: Cuvillier Verlag

Van Laar A and Akça A 2007 *Forest mensuration*, 2nd Edition, New York: Springer

Vanclay JK 1994 *Modelling Forest growth and yield: applications to mixed tropical forests*, Wallingford: CAB International

Vanclay JK and Skovsgaard JP 1997 Evaluating Forest growth models, *Ecological modeling* **98** 1-12

Vose JM and Allen HL 1988 Leaf area, stemwood growth and nutrition relationships in loblolly pine, *Forest Science* **34** 546–563

Wagner C 1928 *Lehrbuch der theoretischen forsteinrichtung*, Berlin: Paul Parey

Wang Jin-Feng, Stein A, Gao Bin-Bo, and Ge Yong 2012 A review of spatial sampling, *Spatial Statistics* doi:10.1016/j.spasta.2012.08.001

Wang M, Rennolls K, and Borders BE 2008 Base-age invariant site index models form a generalized algebraic parameter prediction approach, *Forest Science* **54** 625-632

Weiskittel AR, Hann DW, Kershaw J.A, and Vanclay JK 2011 Forest growth and yield modelling, Oxford: Wiley-Blackwell

Wenk G, Antanaitis V, and Smelko S 1990 *Waldertragslehre*. Berlin: Deutscher Landwirtschaftsverlag

Westoby M 1984 The self-thinning rule, *Advances in Ecological Research* **14** 167-225

Wiant HV Jr, Wood GB, and Furnival GM 1992 Estimating log volume using the centroid position, *Forest Science* **38** 187-191

Wiedemann E 1935 Über die Vereinfachung der Höhenermittlung bei der Vorratsaufnahme, *Mitteilungen aus Forstwirtschaft und Forstwissenschaft* **6** 387-412

Williams MS 1997 A regression technique accounting for heteroscedastic and asymmetric errors, *Journal of Agricultural, Biological and Environmental Statistics* **2** 108-129

Wilson FG 1946 Numerical expression of stocking in terms of height, *Journal of Forestry* 44 758-761

Wood GB, Wiant HV, Loy RJ, and Miles JA 1990 Centroid sampling: a variant of importance sampling for estimating the volume of sample trees of radiata pine, *Forest Ecology and Management* **36** 233-243

Wykoff W 1990 A basal area increment model for individual conifers in northern Rocky Mountains, *Forest Science* **36** 1077-1104

Wykoff WR, Crookston NL, and Stage AR 1982 *User's guide to the stand prognosis model*, USDA Forest Service General Technical Report No. INT-133

Yang RC, Kozak A, and Smith JHG 1978 The potential of Weibull type functions as flexible growth curves, *Canadian Journal of Forest Research* **8** 424-431

Yates F 1936 A new method of arranging variety trials involving large number of varieties, *Journal of Agriculture Science Cambridge* **26** 242-255

Yoda K, Kira T, Ogawa H, and Hozumi H 1963 Self-thinning in overcrowded pure stands under cultivation and natural conditions, *Journal of Biology, Osaka City University* **14** 107-129

Zeide B 1980 Plot size optimization, *Forest Science* **26** 251-257

Zeide B 1993 Analysis of growth equations, *Forest Science* **39** 594-616

Zellner A 1962 An efficient method of estimating seemingly unrelated regressions and tests for aggregation bias, *Journal of American Statistical Association* **57** 348-368

Zellner A and Theil H 1962 Three-stage least squares: Simultaneous estimation of simultaneous equations, *Econometrica* **30** 54-78

Zimmerman DL and Núñez-Antón V 2001 Parametric modelling of growth curve data: An overview (with discussion), *Test* **10** 1–73

Zucchini W and MacDonald IL 1999 Illustrations of the use of pseudo-residuals in assessing the fit of a model, 409-416. *In* H. Friedl, A. Berghold and G. Kauermann [eds.], *Proceedings of the 14th International Workshop on Statistical Modelling*, Graz, Australia